The Complete Guide to

ROOFING & SIDING

Updated 3rd Edition

Choose, Install & Maintain Roofing & Siding Materials

Creative Publishing
international

MINNEAPOLIS, MINNESOTA
www.creativepub.com

Creative Publishing international

Copyright © 2013
Creative Publishing international, Inc.
400 First Avenue North, Suite 400
Minneapolis, Minnesota 55401
1-800-328-0590
www.creativepub.com
All rights reserved

Printed in China

10 9 8 7 6 5 4 3

Library of Congress Cataloging-in-Publication Data

The complete guide to roofing & siding. -- 3rd ed.
   p. cm. -- (Complete guide)
 At head of title: Black & Decker
 Rev. ed. of: Complete guide to roofing, siding & trim / created by: the editors
of Creative Publishing International, Inc., in cooperation with Black & Decker.
2008.
 Includes index.
 Summary: "This updated 3rd edition of Black & Decker's Complete Guide
to Roofing & Siding provides detailed and updated information on modern
materials and techniques for evaluating, installing, and maintaining a wide
variety of roofing and siding materials"--Provided by publisher.
 ISBN 978-1-58923-717-9 (soft cover)
 1. Roofing--Handbooks, manuals, etc. 2. Roofing--Installation--Handbooks,
manuals, etc. 3. Roofs--Maintenance and repair--Handbooks, manuals, etc. 4.
Siding (Building materials)--Handbooks, manuals, etc. I. Creative Publishing
International. II. Complete guide to roofing, siding and trim III. Title: Complete
guide to roofing and siding. IV. Title: Black & Decker

TH2431.C66 2012
695--dc23
                    2011052377

The Complete Guide to Roofing & Siding
Created by: The Editors of Creative Publishing international, Inc., in cooperation with Black & Decker.
Black & Decker® is a trademark of The Black & Decker Corporation and is used under license.

NOTICE TO READERS

Contents

The Complete Guide to Roofing & Siding

Introduction 6

ROOFING 8

Basics of Roofing 10

Evaluating Your Needs 12

Choosing Roofing 14

Estimating Roofing 28

Working Safely 30

Roof Systems 36

Tools & Materials 38

Completing the Tear Off 40

Replacing Sheathing 42

Underlayment 44

Drip Edge 46

Flashing 48

Asphalt Shingles 50

Shingling Over an Old Roof 58

Ridge Vents 60

Cedar Shakes 62

Roll Roofing 68

EPDM Rubber Roofing 74

Raised Seam Metal Roofing 78

Faux Slate 86

Tile Roofing 92

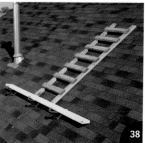

Contents (Cont.)

Inspecting & Repairing a Roof 98

Cleaning Roofs 109

SIDING **110**

Basics of Siding **112**

Choosing Siding 114

Setting Up the Work Site 126

Setting Up Scaffolding 130

Pump-Jack Scaffolding 132

Removing Siding 134

Housewrap 136

Vinyl Siding 138

Lap Siding 146

Wood Shakes & Shingles 156

Plywood Siding 162

Board & Batten Siding 166

Log Cabin Siding 170

Brickmold 176

Finishing Walls with Masonry 178

Brick 180

Mortarless Brick Veneer 184

Cast Veneer Stone 190

Stucco 192

Surface-bonding Cement 198

Painting & Staining Siding 200

Identifying Exterior Paint Problems 202

Preparing to Paint 204

Applying Paint & Primer 208

Using Paint-Spraying Equipment 214

Evaluating Siding & Trim 218

Repairing Siding 220

Repairing Stucco Walls 224

SOFFITS, FASCIA, GUTTERS & VENTS 226

Soffits & Vents 228

Aluminum Soffits 230

Aluminum Fascia 234

Wood Soffits 236

Repairing Wood Fascia & Soffits 238

Gutters 240

Repairing Gutters 246

Conversion Charts 250

Resources 252

Photo Credits 252

Index 253

Introduction

Your home's roofing and siding form a protective envelope against the elements, but their value extends beyond keeping you dry. When selected carefully and installed properly, these systems boost curb appeal—that all-important first impression that adds value to your property, does the neighborhood a favor, and makes your house something to be proud of. So, whether you replace your aging shingles out of necessity or choose to update faded siding to give the place a facelift, you'll enjoy a payback that is both functional and aesthetic. With a little guidance, it's hard to go wrong launching into a roofing or siding project.

You don't necessarily have to call in a contractor to tackle a roof replacement or install new siding. If you are even a moderately experienced do-it-yourselfer and have a spirit of adventure, you have the right stuff to get these projects done effectively. By doing the work yourself, you save a bundle of money—an enticing prospect for any homeowner on a budget. Are you up for the challenge? *The Complete Guide to Roofing & Siding* is the ideal reference to get you started and walk you through every step of the process.

This book is divided into three sections: one on roofing, one on siding, and one on soffits, fascia, vents, and gutters. Each begins with a gallery of product options installed on a variety of homes so you can begin to evaluate the possibilities. An all-important planning section follows, for these are big how-to projects. Doing things in the right order and on a predictable timetable is essential. Each planning section helps you evaluate the challenges of your particular situation and organize the project for the best possible results.

Next, you get a crash course on the latest products available. Options for roofing and siding have never been greater, greener, and more DIY-friendly. With a little time spent in this section of the book, you will go a long way towards getting the look, performance, and value you are hoping for. We also show you how to estimate your material quantities and costs accurately, as well as how to work safely at heights.

As you launch into the actual labor, you'll be guided by thorough step-by-step instructions and insightful photos that show how to do the job right. Roofing projects include asphalt shingles, standing-seam metal roofing, wood shakes and shingles, cement and clay tile, roll roofing, and faux slate. You'll also be introduced to such green approaches as living roofs and photovoltaic shingles. For siding, you'll be guided through lap siding, as well as panel, board and batten, shakes, brick, stucco, and stone veneer. Should you decide the work is beyond your skill set or time availability, this book provides invaluable background for working with a contractor.

We round out each section by showing you how to extend your home's life and maximize your investment with regular maintenance. You'll learn how to carry out essential roofing and flashing repairs. You'll be guided through cleaning, painting, and making siding repairs. Even if these jobs won't be necessary for years to come, you'll be prepared to take action when they are.

Thanks for choosing *The Complete Guide to Roofing & Siding*. We hope it will be the essential reference guide for your exterior home improvement project needs.

ROOFING

Basics of Roofing

The pages that follow will help get you started on the right foot with your roofing project. You'll begin the planning process by evaluating your needs and estimating the material quantities required, the cost, and roughly how long the project will take. Then, you'll take a more in-depth look at the wide variety of product options available.

Safety is an issue with a roofing project. You'll learn important tips for working safely, how to set up ladders, and how to prepare your job site to minimize damage, manage debris, and work efficiently.

The right roofing material for your home is one that looks great, is within your budget, and offers maximum longevity with the least amount of maintenance. Many homeowners opt to simply replace their roofing with the same type—often the most sensible choice. However, if you'd like to change the look of your home or upgrade to a longer lasting material, you'll find plenty of options.

In this chapter:

- Evaluating Your Needs
- Choosing Roofing
- Estimating Roofing
- Working Safely
- Roof Systems
- Tools & Materials
- Completing the Tear Off
- Replacing Sheathing
- Underlayment
- Drip Edge
- Flashing
- Asphalt Shingles

- Shingling Over an Old Roof
- Ridge Vents
- Cedar Shakes
- Roll Roofing
- EPDM Rubber Roofing
- Raised Seam Metal Roofing
- Faux Slate
- Tile Roofing
- Inspecting & Repairing a Roof
- Cleaning Roofs

Evaluating Your Needs

Even if you choose to tackle it yourself, installing a new roof is a major financial investment. You'll find it worth your while to study the many material options. If you do choose to stick with the type of roofing you currently have, you may find an attractive or efficient upgrade in profile, color, or durability.

If you are considering a new type of roof covering, check out the following pages. You'll find a wide range of attractive roofing options that offer some good ways to enhance the look of your home. Bear in mind that while roofing has the all-important function of protecting your home from rain and snow, it has a decorative role as well. A roof packs a lot of visual wallop, making up about a third of a home's façade on average. Neutral colors are a good choice if you are likely to alter the color scheme of your siding and trim. Generally light colored roof coverings are more energy-efficient, and some government programs offer financial incentives for selecting light colored shingles.

The architectural style of your home will seldom limit the choice of roofing material. With the exception of clay tile roofing and some cement tile options that tend to look best on Southwestern- and Mediterranean-style homes, the range of roofing materials can suit almost any home. And roof coverings can be combined. For example, standing-seam metal roofing can top off a bay or bow window for an elegant feature that complements standard asphalt shingles on the main roof deck.

Your roofing serves two fundamental roles: protecting you from the elements, and boosting the attractiveness of your home. For visual variety, you might choose standard asphalt shingles for the bulk of the roof, but add a premium touch like metal roofing above a bay window.

Identifying Roofing Problems

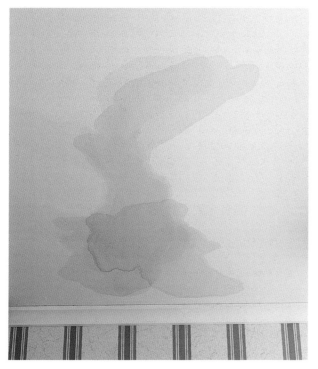

Ceiling stains show up on interior surfaces, but they're usually caused by leaks in the roof.

When shingles start to cup or show signs of widespread damage they need to be replaced.

Loose flashing can be caused by external forces, such as high winds, or by the failure of the sealant or fasteners. The flashing can sometimes be repaired or replaced without replacing the shingles.

Damaged and deteriorated flashing can cause roof leaks. The damaged piece or pieces need to be removed and replaced.

Choosing Roofing

With so many options, choosing a roofing material for your home can seem like a daunting task. Appearance, cost, ease of installation, lifespan, value, and maintenance all play a role.

Style: After you've narrowed your decision down to a type of material, you'll need to choose a style and color. You have quite a few options; even basic asphalt shingles are available in a seemingly endless number of colors. Styles include three-tabs, textured, and scallops.

Architectural features of your house, including the existing materials, can help in your decision making. Stucco siding and clay roofing are a natural fit, especially for a Spanish, Mediterranean, or Southwest styling. Brick siding and metal roofing make a pleasing contrast. Slate (real or faux) is an ideal complement for more traditional homes.

Because roofing lasts for decades, you'll be living with your decision for a long time. Be thorough in looking at different products, weighing their advantages and any disadvantages, before finalizing your decision.

Price: Prices are as varied as the roof-covering products themselves and can differ from region to region. For example, slate is a relatively accessible roofing material in Vermont where it is quarried, but a luxury roofing elsewhere because of its high shipping cost. Cedar shingles and shakes are more expensive in the Northeast than in the Northwest, again due primarily to shipping. However, slate, clay, and copper are generally the most expensive roofing products, followed by concrete tiles, metal, cedar shakes and shingles, organic asphalt, fiberglass asphalt, and roll roofing. If you are married to a particular look but can't afford the real thing, consider less expensive substitutes. The look and features of many premium roofing and siding products are now replicated in other materials. Some metal roofs have the profile of tile, and some asphalt shingles look like the higher-priced wood shakes.

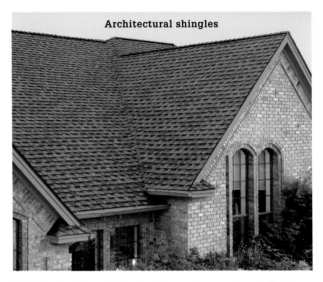

Architectural shingles

Architectural asphalt shingles are the most cost-effective option for adding visual interest to your roof. They are available in a wide range of colors and textures, many of which hide minor installation flaws.

Add Up the Extras ▸

When determining the cost of your project, consider everything you'll need to complete the job, including roofing, flashing, trim, fasteners, and any underlayment necessary for repairs. Don't forget the cost of disposal in the case of a tearoff, and the plastic tarps for protecting plantings. Should you choose to upgrade to a heavier roofing material like concrete tiles, clay, or slate, you may have to bolster rafters as well—a pricey proposition.

ROOFING PRICE COMPARISON
Most Expensive
Slate
Clay
Copper
Mid Range
Concrete tiles
Metal
Cedar shakes/shingles
Self-adhesive roll roofing
Least Expensive
Asphalt
Roll roofing

Durability

In general, maintenance for roofing materials is fairly minimal. On shaded roofs, moss should be kept under control. Any buildup of leaves and fir needles in valleys should be removed to prevent ice dams. Most non-scheduled maintenance is the result of damage to roofing components, such as torn or cracked shingles or loosened metal flashing. Individual asphalt shingles may be damaged by wind or fallen limbs and wooden shakes or shingles may crack and loosen, in which case they can be re-nailed or replaced. Roll roofing can develop blisters or small holes, which can be repaired with roofing cement. Metal, tile, and clay roofs need only minimal maintenance.

Consider the lifespan of the products that you want to install on your roof. Some materials are extremely durable and guaranteed to last 50 years or more, while others may need to be replaced in less than 10 years.

Climate influences product longevity. Long winters with many ice and thaw cycles take their toll on roofs, and so does intense wind. Even intense sunlight is a factor; often a south-facing roof deteriorates more quickly because it bears the full brunt of the sun and its UV radiation for much of the day.

Roof pitch also affects product lifespan. Generally the higher the pitch, the less the likelihood of wind, snow, and ice damage. Flat roofs are more susceptible to damage from falling branches or ice. Standing water is almost inevitable on flat roofs, often turning minor points of weakness into major problems. Keeping nearby trees trimmed back lessens moss buildup and the hazard of damage due to falling limbs.

Wood shakes are available in different durability grades. Top-quality shakes carry 50-year warranties, while the lower-end shakes are rated to last about 30 years. The lifespan of asphalt shingles also varies. They should last a minimum of 20 years, with the thicker, more durable architectural shingles lasting another decade or more. Roll roofing has the shortest life span, lasting between 6 and 12 years. Metal roofs, despite their light weight, are remarkably durable. Warranties vary by product and manufacturer, but warranties for metal roofing of 30 to 50 years or more are typical.

Durability does not necessarily come with a high price tag. Concrete tile roofing is less expensive than slate or real tile, but offers the same benefits of long life, fire resistance, and a premium appearance. Along with concrete, clay tile and slate are the most durable roofing products.

Shakes made of vinyl provide the look of hand-split cedar shakes without the maintenance required for real wood. They're especially suited to damp climates where moss can damage wood shakes.

Asphalt Three-Tab Shingles

Asphalt three-tab roof shingles consist of a sandwich of asphalt and fiberglass or felt layers covered by mineral granules. For several important reasons they've been the dominant roofing option in the United States for more than a century. Weighing 2 to 3 pounds per square foot, asphalt shingles are relatively lightweight compared to slate, clay, or cement tile. Any pitched roof that's suitable for shingling (a 4-in-12 slope or greater) and properly constructed can accept asphalt shingles without further reinforcement.

Asphalt three-tabs are reasonably priced, and you can expect them to last about 20 years, depending of the quality of the shingle and the intensity of the local climate. Installing or repairing asphalt shingles isn't difficult; in fact, it's the most approachable roofing material for do-it-yourselfers or any qualified roofing installer. Ease of installation also helps keep costs down. Once installed, high-quality asphalt singles are relatively maintenance free and offer good fire and wind resistance. They come in a wide range of colors to blend in nicely with any siding and trim color scheme. If you live in a damp or coastal environment, you can also find asphalt shingles impregnated with algaecides to prevent staining.

The most common roof covering by far, asphalt three-tab shingles are inexpensive, lightweight, and available in a wide range of colors and styles. They are also easy to install. They require few tools to install and are do-it-yourself friendly.

Asphalt Shingle Types ▸

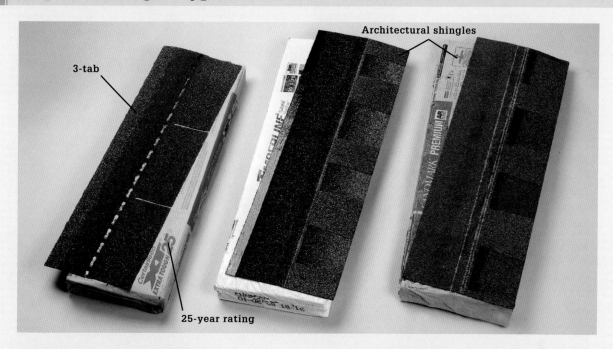

3-tab

Architectural shingles

25-year rating

Shadow-line shingles

4-tabs

Asphalt shingles are usually rated by lifespan, with 20-, 25-, and 40-year ratings the most common (although some now claim to be 50-year shingles). Functionally, these ratings should be used for comparison purposes only. For example, asphalt shingles on a low-pitch roof in a hot climate might fail in as little as 12 years.

The term "multitab shingle" refers to any asphalt shingle manufactured with stamped cutouts to mimic the shapes of slate tile or wood shakes. Multitab cutouts are made and installed in single-thickness, 3-ft. strips, so these tabbed reveals show up. The ubiquitous term for them is "three-tab," but two- and four-tab styles are also available. Generally, the tabs are spaced evenly along each sheet of shingle to provide a uniform appearance and a stepped, brick-laid pattern on the roof. However, some manufacturers also offer styles with shaped corners or randomly spaced tabs trimmed to different heights for a more unique look.

Laminated Asphalt Shingles

Laminated asphalt shingles are an increasingly popular option for new homes or roofing replacements. You may also hear them referred to as "dimensional," "architectural," or "multithickness" shingles. Essentially, laminated asphalt shingles have the same material composition as multitab cutout shingles—a sandwich of asphalt, base sheeting of felt or fiberglass, and granular coatings. However, the important difference is that instead of a single shingle layer, two layers are bonded together to create a three-dimensional appearance. The top layer has wide, randomly sized notches, and it overlays a staggered, unnotched bottom layer. The effect of this lamination treatment helps mimic the natural inconsistencies of a cedar shake or slate roof.

Laminated shingles are no more difficult to install than three-tab shingles, and the same installation methods apply. The random pattern can even reduce overall installation waste because the pattern is more visually forgiving than the uniform design of three-tab shingles. Sections of shingle that would otherwise interrupt a three-tab pattern can still be used in a laminated shingle installation.

The added shingle layer makes laminated shingles heavier than standard three-tab shingles, but the weight difference isn't critical; any pitched roof can accept them without special structural reinforcement. Laminated shingles are also thicker than multitab shingles, which gives them improved wind resistance and durability. As you might expect, they're more expensive than three-tab shingles. The added thickness enables manufacturers to offer longer 30- to 50-year warranties, which can make them a better investment than three-tab shingles over the life of your roof.

Architectural shingles (also called laminated and dimensional) are installed very similarly to regular three-tab asphalt shingles, but they offer a more sophisticated appearance with greater visual depth (inset). They also come with longer warranties than comparable three-tab shingles. They're easy for homeowners to install, only marginally more difficult than standard three-tabs, and require only basic tools.

Asphalt Roll Roofing

For low-pitched roofs, or in cases where budgetary concerns are more important than appearance, rolled roofing might be the perfect choice for your situation. Roll roofing is also a good option if you need to cover a roof for an extended period but plan to install the final roofing at a later date.

Think of asphalt-and-granule roll roofing as continuous strips of asphalt shingles without the tab cutouts. Roll asphalt roofing is installed in overlapping layers just like asphalt shingles, but the amount of exposure from one layer to the next is much greater than it is with roll shingles. Depending on the product, asphalt roll roofing will either be entirely covered with granules and installed to provide a single layer of coverage, or only half the sheet will be covered in a granule layer. The uncoated half (called a selvedge edge) and half of the shingle layer above it are covered with roofing cement for an improved seal.

The wider overlap provided by the selvedge edge lends a double-coverage effect. (Double-coverage roofing is rapidly becoming rare.)

Single-coverage roll roofing may also have a selvedge edge for cement, but it will be only a few inches wide. Single- and double-coverage styles are secured with roofing nails around the edges. Both types are suitable for do-it-yourself roofing.

New self-adhesive products have lifted this type of roofing out of the bargain basement. Made of two layers—a base layer covered by a granule-coated top layer—these products install much like ice and water shield and offer better appearance and increased longevity. The adhesive is fierce, all but eliminating any tendency to blister. Because of the adhesion and two-layered installation, manufacturers warranty the material for 15 years. Most manufacturers of standard roll roofing offer no warranty at all.

Roll roofing is essentially building paper with a granular surface coating. Single-coverage roll roofing is typically installed in a perimeter bond application with an overlap of about 6". Self-adhesive ice and water shield is a roll roof product that is installed under the first few courses of shingles to prevent leaks from ice dams. Self-adhesive roll roofing has two layers and offers better longevity than typical roll roofing.

EPDM Rubber Roofing

Rubber membrane roofing, also known as ethylene propylene diene monomer (EPDM) roofing, is installed in wide sheets using a specialized latex bonding adhesive, but it is not nailed in place. Rubber roofs for home use are almost always fully bonded to the roof deck with cement. The absence of nails make it an even better choice for low-pitch roofs that are susceptible to leaks. You may choose to install a rubber membrane on a pitched roof simply because it is seamless and easy to install.

The membrane is available in 10- or 20-ft.-wide rolls to help reduce the total number of seams. Common thicknesses are 45 and 60 mil (.04 and .06 in.). In recent years, EPDM membrane has become more do-it-yourself friendly, and it's available to consumers through roofing suppliers and some home centers. It is comparably priced to quality asphalt shingles. For heat reflection and an added layer of protection, some homeowners opt to add gravel ballast or, if the structure can bear it, concrete pavers.

Rubber membrane roofs used to be installed almost exclusively in commercial situations, but DIY-friendly versions have become common in recent years. They're an excellent choice for flat or near-flat roofs. The rubber material is sold in prepackaged rolls (50 and 100 ft. are common) or you can buy it by the lineal foot from some roofing material suppliers.

Slate Shingles

Slate roofing has been around for centuries, and it's one of the most weather-resistant and beautiful options you can put on your roof. It is quarried and cut into thin individual shingles and installed with nails. Natural variations in color and texture give slate an organic quality that is unmatched by synthetic roofing products.

However, despite its visual appeal and long-lasting performance—more than 100 years in many cases—a number of important factors may make slate unfeasible for your home. For one, slate is heavy, weighing about twice as much as asphalt singles per square foot. It's also about triple the cost of premium asphalt shingles. A conventionally framed or truss roof may require additional framing before it's suitable for slate shingles. Check with a structural engineer to see if your home's roof will support slate.

Installing slate shingles involves the use of specialized cutting tools and skills, which makes it extremely complicated to install for do-it-yourselfers as well as many roofers. Once installed, slate's durability to the elements doesn't extend to foot traffic. The shingles are brittle and can break if they are stepped on. Replacing broken shingles involves installing extra staging and ladders to prevent further damage, and removing individual shingles is a complicated process.

Slate roofs are expensive and not DIY-friendly, but for overall attractiveness and durability, they're hard to beat. Due its longevity, slate requires premium flashing materials and careful installation.

Clay Tile

Clay tile roofs are common on southern and coastal homes, where intense heat and high winds are a concern. Since clay is a manufactured product, roofing tiles are available in a wide range of shapes, sizes, and colors. Clay tiles offer excellent durability and fire resistance. However, clay is slightly heavier than slate, and installing it over typical roof framing may require adding structural reinforcements. Of course, this adds to the project costs.

Depending on the region you live in, there may be many qualified roofers who can install it properly. Working with clay tile requires cutting with a diamond-blade saw, and it may not be suitable for complex roof designs. It is also relatively fragile and can not be walked on, so repairs can be difficult. The installed cost of clay tiles is comparable to slate. On all but the most basic roof designs, clay tile is not do-it-yourself friendly.

Clay tile roofs have a distinctive regional appearance, but they can be installed in practically any climate. They're heavy and relatively expensive. Most clay tiles have a half-pipe shape and terra-cotta color, but with a little research you can find a rather wide range of colors and styles. Due to its weight and unique method of installation, clay tile is a challenging material for a do-it-yourselfer.

Concrete Tile

Over the past few decades, concrete has continued to gain momentum in the roofing field. It weathers well; offers excellent wind, hail, and fire resistance; and installs similarly to clay tile and slate. Concrete is typically formed into flat shakes instead of contoured tiles, like clay, but you can special-order contoured shapes made from cement. Some manufacturers offer concrete shakes that mimic the color and texture of wood.

As you might expect, concrete shakes are heavy. They're slightly heavier than slate and about the same weight as clay tile. However, there are also concrete shakes that are fortified with wood fibers and polymers to cut down on overall weight without compromising durability. Compared with clay or slate, concrete offers similar performance characteristics with a lower overall installation cost: it's about half the price of slate or clay.

Fiber cement is growing quickly as a concrete roofing tile material, much as it is for lap siding. It offers a singular package of durability, low maintenance, beauty, and reasonable cost. Cement tiles and fiber-cement tiles usually have a simulated wood grain appearance to resemble wood shingles, but concrete tiles can be ordered in many other configurations that are less well known. Though heavy and challenging to cut, concrete tiles are straightforward to install.

Metal Roofing

Metal roofing has proven its durability as a residential roofing material for centuries in Europe, but until recently it was more common to see a metal roof on an agricultural or commercial building than on a home here in the States. That trend is quickly changing. Advances in metal-forming techniques and improvements in coatings have created a wide variety of styles and colors to choose from, making metal roofing a more enticing option for homeowners. In fact, metal roofing is the fastest growing segment of the residential roofing products market for several reasons. For one, metal is the lightest-weight roofing material made. Because metal roofing weighs less than most other roofing materials, any standard-framed roof can support it. Provided the shingles are in sound condition and local building codes permit it, metal roofing can even be installed over a layer or two of asphalt shingles. This saves on the cost and effort involved with a tearoff.

Metal offers excellent wind and fire resistance, and improvements in rust-inhibitive coatings make it weather well for many decades. It resists peeling, chalking, and fading from UV light. Metal roofs are quite common in coastal areas that are subjected to tropical storms and high winds. They also perform favorably when subjected to heavy snowfall or ice accumulations.

Standing-seam steel roofs are lightweight, very durable, and only moderately expensive. Typically, they are fabricated on-site by pros, but some prefabricated panel systems are DIY friendly. Installing metal roofing requires metal cutting skills and the selection of a product that doesn't require special installation equipment.

Metal shingles offer color, dimensionality, and long life. They install in panels one course deep and 2- to 4-ft. long. Some types are granular coated, looking much like concrete tiles but without the added weight.

Metal roofing shaped like low-profiled tile comes not just in terra cotta, but red, blue, gray, green, and brown—a good option if you plan to make your roof a design feature of your home.

Unlike wood or asphalt roofing, metal will not rot, crack, or promote algae growth, so it is largely maintenance free. Light-colored metal roofs, which reflect sunlight instead of absorb it, offer energy-saving benefits as well. Attics stay cooler in summer months, which reduces energy costs. Given its durability, warranties in the 30- to 50-year range are common for professionally installed metal roofing systems. In terms of cost, metal is more expensive than premium asphalt shingles but cheaper than slate or clay.

Residential metal roofing is available in steel, aluminum, or copper. It is embossed in several surface textures to simulate clay tiles, wood shakes, or asphalt shingles. Some products are coated with granules to enhance their texture. Embossed metal roofing comes in either horizontal panels with several shingles formed on each panel or as individual shake or tile shapes. Each piece has flanges that clip to the top and bottom of adjacent pieces or nail-down clips along the edges. Smooth panels with standing seams crimped or clipped together along both long edges are the most common style of metal roofing. The metal panels can be formed on location with common metalworking tools. They're also available in prefabricated panels. Copper is most often installed in panel form and left unfinished to weather to a green patina. It is the most expensive option. Steel or aluminum roofing come in a spectrum of colors to suit any siding color scheme.

There are a few drawbacks to metal roofing worth noting. It is slippery when wet and it can be dented or scratched. Metal can be difficult to adapt to complex roof styles. Eventually steel roofing will need to be repainted to prevent corrosion.

Although it shouldn't be difficult to find qualified installers in most regions, you need special metalworking skills to tackle most residential metal roofing products yourself.

Corrugated Nonmetallic Panels ▸

Prefabricated panels made of metal, fiberglass, or clear polycarbonate come in standard widths and lengths and are usually installed over a system of purlins. Historically, they have been used most often to roof outbuildings and shelters, but the metal versions are becoming more popular for whole houses and room additions.

Environmentally Friendly Roofs

A little-known but new breed of environmentally friendly roofing products is now available for residential applications. If you like the look of slate, tile, or wood shakes, you can now buy them made from blends of polymers, sawdust, rubber, vinyl, and fiberglass. These composite shingles are quite flexible, so they can withstand high winds, hail damage, or foot traffic without splitting, rotting, or breaking. Fire retarders and UV inhibitors are added to make them as durable as conventional roofing options. In polymer-based shingles, the color is blended through the material, so scratches won't show and peeling isn't a concern. Some of these alternative roofing products are made from recycled materials, such as used car tires or post-industrial waste, and the roofing may be entirely recyclable when it wears out.

Polymer or rubber composite roofing materials are heavier than metal but they are about the same weight as laminated asphalt shingles. Most come with 50-year warranties that can be transferred from one homeowner to the next. They install easily with nails over typical felt underlayment, just like asphalt shingles. You may be able to install a composite roof yourself, depending on the product.

Fly ash

Slag

Green products on the roof, as in every other part of the house, are beginning to flood the market. This welcome occurrence is creating a whole new category of options for homeowners with pro-environment priorities. Most of the newer products, like those seen above, are made with recycled materials using composite technologies. Others, however, feature slightly older technology (see Living Roofs, next page). Local tile materials (from within a 500-mile project radius) manufactured with indigenous materials and postindustrial recycled content, including fly ash and slag, are shown here. These criteria contribute toward third party certification and possible leadership in energy and environmental design (LEED) credits.

Green Roof Options ▸

LIVING ROOFS

Once a mainstay on the Great Plains where no other roofing material was available and only sod made sense, living roofs made of plants and planting media are well established in commercial construction and used increasingly in residential architecture. Modular systems, installed over an waterproof membrane like EPDM, have taken much of the risk out of living roofs. Here are the benefits:

- Properly maintained, a living roof can last 50 years.
- Living roofs reduce storm water runoff, cutting waterway pollution and energy demands on sewage treatment plants.
- A living roof adds insulation, reducing heating costs and cutting cooling costs as much as 25 percent.
- Living roofs improve air quality. A two-car garage with a living roof will create enough oxygen annually to supply 25 people for a year.
- A living roof provides habitat for wildlife.
- Green roofs can host some types of vegetables for a beautiful overhead addition.

Whether you choose to add a living roof to your home or experiment with the approach on your garage, there are three primary methods of achieving a living roof. The first is to hire a design-and-build firm that specializes in the installation of living roofs; the second is to build one yourself using a proprietary kit-type system; last, you can make one mostly from products you can purchase at a home center. Here are some planning tips to consider:

- Avoid slopes greater than 2-in-12. It is possible to create a living roof on a steeper pitch, but it is much more complicated.
- Consult with your local building department and a structural engineer to make sure your roof framing is adequate to support the weight of a living roof. Depending on the growing medium and drainage method used, a living roof can add up to 100 pounds per square foot.
- Unless your current roof covering is a waterproof membrane such as EPDM rubber (the preferred roofing) or bonded PVC sheets, you will need to reroof.
- Are you comfortable working at heights? You will need to do occasional weeding and replanting, and unless you choose drought-resistant plants, may need to water as well.

PHOTOVOLTAIC SHINGLES

If you aren't ready for something photosynthetic on your roof, how about something photovoltaic? Until very recently, if you wanted to put your roof to work generating power, you had to tolerate clunky—and expensive—solar panels. Other options like flexible photovoltaic strips on flat or metal roofs and solar panels roughly as thick as concrete tiles were good first steps, but new integrated photovoltaic shingles are truly suited to a roof on any suburban cul-de-sac. Melded into asphalt roofing (they almost disappear on a black roof), photovoltaic shingles protect you from the elements while feeding electricity into your house. Thin and flexible, much like asphalt three-tabs, they nail in place, each subsequent course covering the nail heads of the previous course. As they are installed they link together with built-in plugs. The shingles are tough enough to withstand a hammer drop, as well as rough weather.

The estimated $25,000 for product and installation is daunting, but is offset by 30 percent federal tax credits, as well as state and utility incentives, all of which combine to bring the price tag down to about $10,000. Once up and running, the shingles can potentially produce half the power a home needs. Excess power feeds back into the grid.

These are the early days for this technology, and you may want to wait until this approach has proven itself. But for simplicity and integrated appearance, this system is well worth watching.

Eco-friendliness and natural insulation help offset the cost of a living roof. If your roof slope is slight and its framing beefy enough to stand the extra weight, you can choose from several manufactured systems to add overhead foliage to your home.

Estimating Roofing

Roofing materials are ordered in squares, with one square equaling 100 square feet. To determine how many squares are needed, first figure out the square footage of your roof. The easiest way to make this calculation is to multiply the length by the width of each section of roof, and then add the numbers together.

For steep roofs and those with complex designs, do your measuring from the ground and multiply by a number based on the slope of your roof. Measure the length and width of your house, include the overhangs, then multiply the numbers together to determine the overall square footage. Using the chart at the lower right, multiply the square footage by a number based on the roof's slope. Add 10 percent for waste, then divide the total square footage by 100 to determine the number of squares you need. Don't spend time calculating and subtracting the areas that won't be covered, such as skylights and chimneys. They're usually small enough that they don't impact the number of squares you need. Besides, it's good to have extra materials for waste, mistakes, and later repairs.

To determine how much flashing you'll need, measure the length of the valley to figure valley flashing, the lengths of the eaves and rakes to figure drip edge, and the number and size of vent pipes to figure vent flashing.

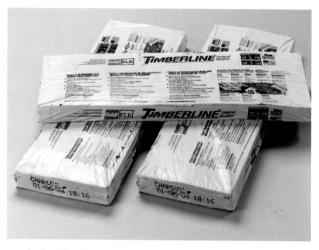

Asphalt shingles come in packaged bundles weighing around 65 pounds each. For typical three-tab shingles, three bundles will cover one square (100 sq. ft.) of roof.

Calculate the roof's area by multiplying the height of the roof by the width. Do this for each section, then add the totals together. Divide that number by 100, add 10 percent for waste, and that's the number of squares of roofing materials you need.

CONVERSION CHART

Slope	Multiply by	Slope	Multiply by
2-in-12	1.02	8-in-12	1.20
3-in-12	1.03	9-in-12	1.25
4-in-12	1.06	10-in-12	1.30
5-in-12	1.08	11-in-12	1.36
6-in-12	1.12	12-in-12	1.41
7-in-12	1.16		

Measuring and Planning

Calculate the slope of your roof before beginning any roofing project. Roof slope is defined as the number of inches the roof rises for each 12" of horizontal extension (called the "run"). For example, the roof shown above has a 5-in-12 slope: it rises 5" in 12" of run. Knowing the slope is important when selecting materials and gauging the difficulty of working on the roof. To ensure safe footing, install temporary roof jacks if the slope is 7-in-12 or steeper. Roofs with a slope of 3-in-12 or less require a fully bonded covering to protect against the effects of pooling water.

How to Measure a Slope ▶

Hold a carpenter's square against the roofline, with the top arm horizontal (check it with a level). Position the square so it intersects the roof at the 12" mark. On the vertical arm, measure down from the top to the point of intersection to find the rise.

Protect against damage from falling materials when working on the roof. Hang tarps over the sides of the house, and lean plywood against the house to protect the vegetation and siding.

ESTIMATING TIME REQUIREMENTS

Task	Time Required	× Amount	= Total Time
Tear-off	1 hr./square*		
Install felt paper	30 min./square		
Apply shingles:			
Flat run	2 hrs./square**		
Ridges, hips	30 min./10 ft.		
Dormers	add 1 hour each		
Flashing:			
Chimneys	2 hrs. ea.		
Vent pipes	30 min. ea.		
Valleys	30 min./10 ft.		
Roof vents	30 min. ea.		
Skylights	2 hrs. ea.		
Drip edge	30 min./20 ft.		

TOTAL TIME FOR PROJECT

Note: All time estimates are based on one worker. Reduce time by 40% if you have a helper.

*One square=100 sq. ft.

**Include area of dormer surface in "flat run" estimate

Working Safely

Working on the exterior of a house presents challenges not faced in the interior, such as dealing with the weather, working at heights, and staying clear of power lines. By taking a few commonsense safety precautions, you can perform exterior work safely.

Dress appropriately for the job and weather. Avoid working in extreme temperatures, hot or cold, and never work outdoors during a storm or high winds.

Work with a helper whenever possible—especially when working at heights. If you must work alone, tell a family member or friend so the person can check in with you periodically. Keep your cell phone handy at all times.

Don't use tools or work at heights after consuming alcohol. If you're taking medications, read the label and follow the recommendations regarding the use of tools and equipment.

When using ladders, extend the top of the ladder 3 feet above the roof edge for greater stability. Climb on and off the ladder at a point as close to the ground as possible. Use caution and keep your center of gravity low when moving from a ladder onto a roof. Keep your hips between the side rails when reaching over the side of a ladder, and be careful not to extend yourself too far or it could throw off your balance. Move the ladder as often as necessary to avoid overreaching. Finally, don't exceed the workload rating for your ladder. Read and follow the load limits and safety instructions listed on the label.

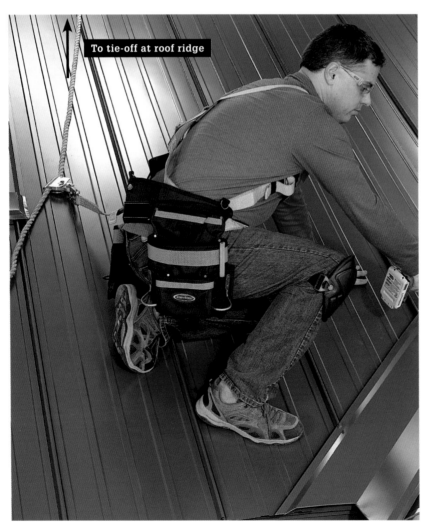

To tie-off at roof ridge

Wear appropriate clothing and safety equipment whenever working at heights. Eye protection and hearing protection are important when using power tools or pneumatic tools. And if you'll be climbing on a roof, wear tennis shoes or any sturdy shoe with a soft sole designed for gripping. When roofing, always avoid hard-soled shoes or boots, which can damage shingles and are prone to slipping. On sloped roofs, fall-arresting gear is always recommended. The rig seen here (see Resources, page 252) is affordable for homeowner use. You can also rent fall-arresting gear at many rental centers.

Use a GFCI extension cord when working outdoors. GFCIs (ground-fault circuit-interrupters) shut off power if a ground fault occurs in the electrical circuit.

Enlist a helper to raise an extension ladder. A reliable rule of thumb for the correct distance from the ladder feet to the bracing wall is ¼ of the height of the ladder's extension (if the ladder is extended 12 feet, position the feet of the ladder 3 feet from the wall). Once in position, and with a helper bracing the ladder feet, extend the ladder to the required height and then walk it upright until it rests against the house.

Use fiberglass ladders when working near power cables. Exercise extreme caution around these cables, and only work near them when absolutely necessary.

Never climb a ladder with a loaded air nailer attached to a pressurized air hose. Even with trigger safeties, air guns pose a serious danger to the operator as well as anyone who may be standing near the ladder. Connect the gun to the hose only after you are off the ladder safely.

Attach a ladder jack to your ladder by slipping the rung mounts over the ladder rungs. Level the platform body arm, and lock it into place.

Set the plank in place on the platform arm. Adjust the arm's end stop to hold the plank in place.

Stabilize your ladder with stakes driven into the ground, behind each ladder foot. Install sturdy blocking under the legs of the ladder if the ground is uneven.

Fly hooks

CAUTION: Use care when approaching transition steps between flies.

Make sure both fly hooks are secure before climbing an extension ladder. The open ends of the hooks should grip a rung on the lower fly extension.

Fall-Arresting Gear

Even if you consider yourself dexterous and are comfortable working in high places, all it takes is one misstep on a roof to lead to a tragic fall. Despite the fact that many professional roofers never don safety harnesses, you should seriously consider investing in personal fall-arresting gear if you plan to reroof your home.

Fall-arresting gear consists of several components. Wear a webbed body harness that spreads the impact of a fall over your shoulders, thighs, and back to reduce injury. Harnesses are made to fit average adult builds. The harness connects to a shock absorber and a lanyard around 6 feet in length. A self-locking, rope-grab mechanism attaches the lanyard to a lifeline that must be fastened securely to a ridge anchor screwed to roof framing. In the event that you slip or fall, the rope grab will limit your fall to the length of the lanyard because it will not move down the lifeline unless you override the locking mechanism by hand.

Many rental centers carry fall-arresting gear, or you can buy a complete system for less than $100 (see Resources, page 252). When compared with the loss of life or limb, however, your real investment is small. Better yet, you'll have it available any time you need to get on the roof for cleaning tasks or to make repairs.

Tools & Materials ▸

Pry bar	Rope grab
Drill/driver	Synthetic fiber lifeline
Harness	Ridge anchor
Lanyard	

A metal ridge anchor must be secured with screws to the roof framing. Follow the manufacturer's recommendations for proper screw sizing, and make sure your attachment points go beyond the roof sheathing into the roof trusses or rafters.

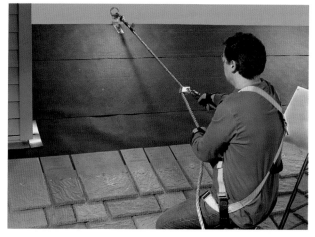

The rope-grab mechanism allows you to move up a roof along the lifeline without interference. To move down the roof, you'll need to override the grab by hand. As soon as you release it, the lock engages again.

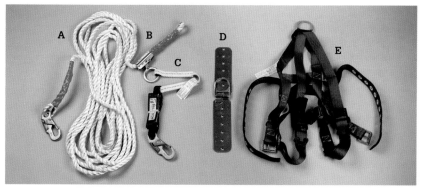

Personal fall-arresting gear consists of a lifeline (A) with mechanical rope grab (B) and lanyard (C); a metal ridge anchor (D); and a body harness (E).

Lifting & Staging Shingles

Carrying bundles of asphalt shingles up a ladder and onto a roof deck is a grueling job, especially if you have as many as 50 bundles to haul. With each bundle weighing 65 to 75 pounds, you need to be concerned about two issues: your safety and how to stage the material for efficient installation. The easiest solution is to simply have your roofing supplier unload the shingles directly to your roof with a boom truck or conveyor belt. This will cost a nominal fee, but the physical exertion you'll avoid may be well worth the cost.

If you must manually unload the bundles from your supplier while on the roof, place the first bundle or two flat on the roof to serve as a base, then stack subsequent bundles partially on the first two and partially on the roof deck. Limit your stack sizes to a dozen bundles or less to help distribute the weight. If possible, place the bundles at a point where two roofs intersect for added stability. Distribute your stacks evenly along the length of the roof ridge so they are readily available wherever you're working as the job progresses. Since the ridge is the last area shingled, most of your supply should be used and out of the way by the time you reach the peak.

For smaller jobs, you may elect to simply carry the bundles up to the roof yourself. In this situation, wear a back support brace to prevent back strain, and carry one bundle at a time over your shoulder so you can keep one hand on a ladder rung at all times. Hand off each bundle to a helper waiting for you on the roof. Switch jobs before you get tired to conserve your energy and share the really hard work.

How to Stage Shingles

When you must carry bundles up a ladder, wear a lumbar support brace to prevent back strain. Take your time carrying each bundle, and rest often. Make sure the bundle is well balanced before you ascend the ladder. If you feel the weight of the bundle start to shift, drop the bundle immediately for your safety. Have a helper waiting for you when you reach the roof deck so you don't have to unload bundles near the eave or risk a fall.

Build stable stacks of a dozen bundles of shingles near the roof ridge. Create stacks by placing two bundles flat on the roof, then adding more bundles that straddle both the first two and the roof deck. When possible, stack bundles where two roof areas meet.

Roof Jacks ▸

Sure footing isn't an issue when you're working on a low-pitched roof, but it becomes a real safety concern for roofs with 7-in-12 or steeper pitches. In these situations, you need to install roof jacks to create a stable work area and navigate the roof safely. Roof jacks are steel braces that nail temporarily to roof decking to support a 2 × 8 or 2 × 10 perch. In addition to improving your footing, roof jacks also provide a flatter surface to stand on, which can help reduce ankle strain. Roof jacks should be installed every 4 feet of plank length with 16d nails. They're inexpensive and available wherever roofing products are sold.

Tools & Materials ▸

Pry bar
Hammer
16d nails
Roof jacks
2 × 8 or 2 × 10 lumber

16d nails

Roof jacks are steel braces that are nailed to the roof deck. Installed in pairs, they support a dimensional board (usually a 2 × 8) to create a sturdy work platform on a sloped roof.

How to Install Roof Jacks

Nail roof jacks to the roof at the fourth or fifth course. Drive 16d nails into the overlap, or dead area, where they won't be exposed. Install one jack every 4 ft., with a 6" to 12" overhang at the ends of the boards.

Shingle over the tops of the roof jacks. Rest a 2 × 8 or 2 × 10 board on the jacks. Fasten the board with a nail driven through the hole in the lip of each roof jack.

When the project is complete, remove the boards and jacks. Position the end of a flat pry bar over each nail and drive in the nail by rapping the shank with a hammer.

Roof Systems

The elements of a roof system work together to provide shelter, drainage, and ventilation. The roof covering is composed of sheathing, felt paper, and shingles. Metal flashing is attached in valleys and around chimneys, vent pipes, and other roof elements to seal out water. Soffits cover and protect the eaves area below the roof overhang. Fascia, usually attached at the ends of the rafters, supports soffit panels as well as a gutter and downspout system. Soffit vents and roof vents keep fresh air circulating throughout the roof system.

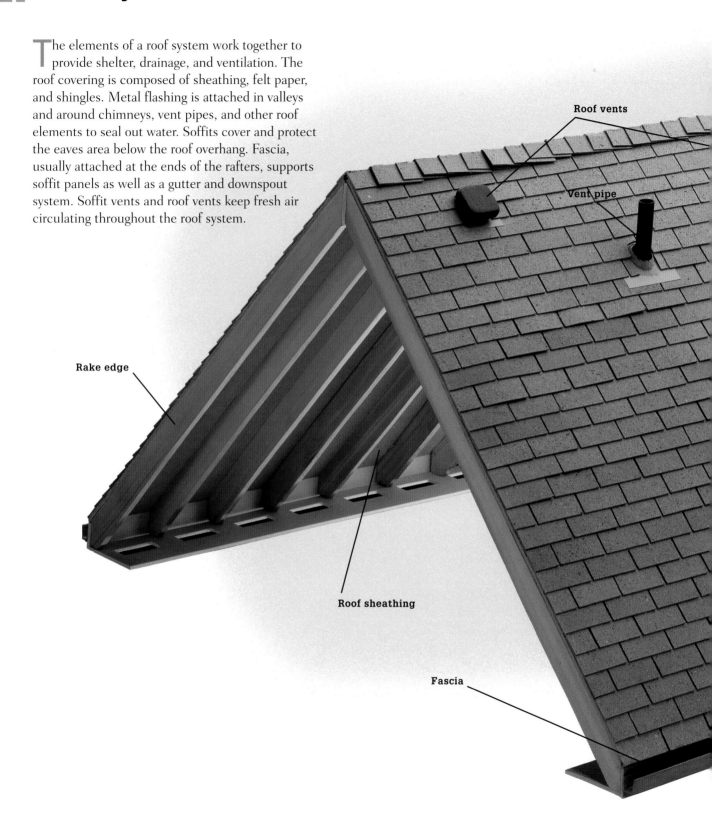

Roof vents

Vent pipe

Rake edge

Roof sheathing

Fascia

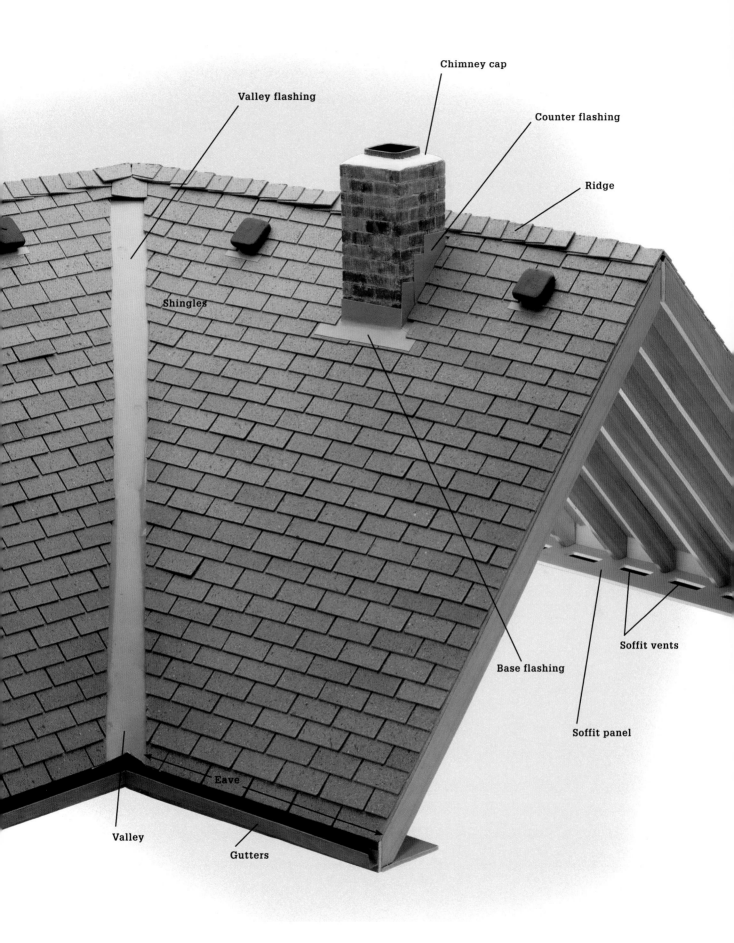

Chimney cap

Counter flashing

Ridge

Valley flashing

Shingles

Base flashing

Soffit vents

Soffit panel

Eave

Valley

Gutters

Tools & Materials

Roofing is tough work, hard on the back, hands, and upper body. Make the job as easy as possible by gathering the right tools and equipment before you begin. Doing so will make the work go faster and spare you physical wear and tear. Many of the necessary tools will prove useful for other DIY projects.

Some of these tools, such as a pneumatic nailer or a roofer's hatchet, are specific to roofing projects. If you don't have them and don't want to buy them, they are rentable. Here are the essential tools you'll need for most roofing jobs.

For more secure footing, fashion a roofing ladder by nailing wood strips across a pair of 2 × 4s. Secure the ladder to the roof jacks, and use it to maintain your footing.

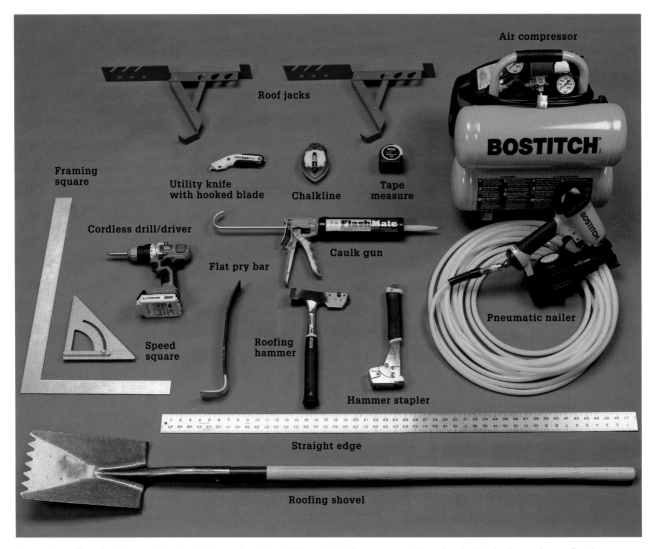

Several roofing tools may already be in your toolchest. Other tools, like a pneumatic nailer, ladder jacks, and a roofing hammer, are wise additions to your DIY arsenal.

Rolled flashing

Drip edge

Preformed valley flashing

Aviation snips

Vent pipe flashing

Step flashing blanks

Skylight flashing kit (partial)

Roof flashing can be hand cut or purchased in preformed shapes and sizes. Long pieces of valley flashing, base flashing, top saddles, and other nonstandard pieces can be cut from rolled flashing material using aviation snips. Step flashing blanks can be bought in standard sizes and bent to fit. Drip edge and vent pipe flashing are available preformed. Skylight flashing usually comes as a kit with the window. Complicated flashings, such as chimney crickets, can be custom fabricated by a metalworker.

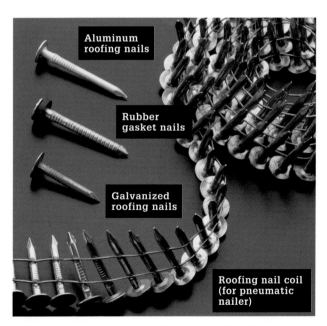

Aluminum roofing nails

Rubber gasket nails

Galvanized roofing nails

Roofing nail coil (for pneumatic nailer)

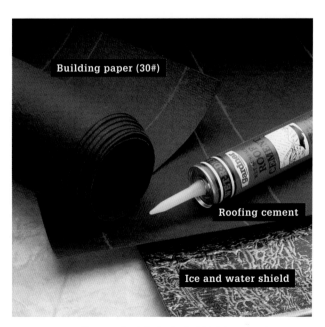

Building paper (30#)

Roofing cement

Ice and water shield

Different fasteners are specially developed for different jobs. Use galvanized roofing nails to hand nail shingles; use aluminum nails for aluminum flashing; use rubber gasket nails for galvanized metal flashing; and use nail coils for pneumatic nailers.

Common roofing materials include 30# felt paper for use as underlayment; ice and water shield for use as underlayment in cold climates; and tubes of roofing cement for sealing small holes and adhering shingles or flashing.

Completing the Tear Off

Removing shingles, commonly referred to in the roofing trade as the tear off, can be done rather quickly. This makes it one of the more satisfying parts of a reshingling project. If you can't reshingle your entire roof in one day, tear off one section of roofing at a time, roof that section, then move on to the next part of the roof.

The tear off produces a lot of waste. A few preparatory steps make cleanup much easier (see page 29). Lay tarps on the ground and lean sheets of plywood against the house to protect shrubbery and the siding.

If renting a dumpster isn't practical or you can't get one close to the roof, set wheelbarrows on tarps as an alternative for catching debris. However, you'll still be responsible for disposing of the old roofing materials, which will probably require several trips to the landfill. To work efficiently, have another person deal with the debris on the ground as you work on the roof. Always wear appropriate safety gear.

Tools & Materials ▸

Wheelbarrow or dumpster	Release magnet
Hammer	Rake
Chisel	Tin snips
Pry bar	Reciprocating saw
Utility knife	Drill
Roofing shovel or pitchfork	Protective gear
Broom	Tarps

Rent a dumpster from a waste disposal company or your local waste management department. If you are re-roofing, position the dumpster directly below the roof edge, so when you're tearing off the old roofing materials, the debris can be dumped from the roof directly into the dumpster.

How to Tear Off Old Shingles

1

Remove the ridge cap using a flat pry bar. Pry up the cap shingles at the nail locations.

2

Working downward from the peak, tear off the building paper and old shingles with a roofing shovel or pitchfork.

3

Remove flashing by cutting and prying. Avoid damaging pipes and vents. You may be able to salvage longer flashing pieces, such as chimney saddles and crickets.

4

After removing the shingles, building paper, and flashing from the entire tear-off section, pry out any remaining nails and sweep the roof. In some cases it is easier to drive nail heads flush with the roof surface.

5

If an unexpected delay keeps you from finishing a section before nightfall, cover any unshingled sections using tarps weighted down with shingle bundles.

Replacing Sheathing

Once the tear off is complete, inspect the roof decking for damage. If there are any soft spots on the roof, or if a portion of the sheathing is damaged, it'll need to be replaced. Most older roofs are constructed with board sheathing, usually 1 × 6s, while newer roofs typically use 4 × 8-ft. sheets of plywood or oriented strandboard (OSB). Even if your roof has board sheathing, you can make the repairs with plywood, as we're doing here. Make sure the plywood is the same thickness as your current sheathing and rated for exterior use.

Before cutting into your roof, check under the sheathing for wires. There may be telephone wires or television cable hidden in the roof, and you don't want to cut through them. Avoid walking on the damaged sheathing. If you have access from the underside, it may be safer to remove the sheathing from below.

Damage to the roof sheathing normally occurs because there is a violation of the roof seal, typically occurring around a chimney, roof vent, skylight or another flashed object. If you will be reflashing the roof make sure you don't repeat any mistakes. If you are making only localized repairs, be sure that you identify and correct the source of the moisture that's caused the deterioration. If the damage is located near the eave and is not caused by a flashing problem or a leak in the roof covering, it is probably caused by an ice dam (see page 98).

Tools & Materials ▸

Circular saw	Sheathing
Reciprocating saw	2 × 4 nailing strips
Tape measure	3", 2½" deck screws
Chalkline	Plywood
Flat pry bar	8d ring-shank
Drill	siding nails

Inspect for damaged sheathing after tear off is completed. Replace damaged roof deck, making sure the new seams fall over rafters. Also replace trim boards in the repair area if they have become damaged.

How to Replace Damaged Sheathing

Use a reciprocating saw to cut next to the rafters in an area that extends well beyond the damaged area. Pry out the damaged sections using a pry bar.

Attach 2 × 4 nailing strips to the inside edges of the rafters using 3" deck screws

Use exterior grade plywood or OSB to make a patch. Measure the cutout area, allow for a ⅛" gap on all sides for expansion, and cut the patch to size. Attach the patch to the rafters and nailing strips using 2½" deck screws or 8d ring-shank siding nails.

Option: If your existing roof deck is made of boards (1 × 6 was common before plywood took over the market), it is perfectly acceptable to use plywood when replacing a section of the deck. The plywood should be the same thickness as the boards, generally ¾".

Underlayment

Building paper, also called felt paper or tar paper, is installed on roof decks as insurance in case leaks develop in shingles or flashing. It's sold in several weights, but heavier 30# paper is a good choice for use under shingles, and may be required by code.

In cold climates, codes often require an additional underlayment called "ice and water shield" or "ice guard" that's used instead of standard building paper for the first one or two course of underlayment and in valleys. In cold climates, apply as many courses of ice and water shield as it takes to cover 36 inches past the roof overhang. An adhesive membrane, the ice guard bonds with the roof sheathing and seals around nails to create a barrier against water backing up from ice dams.

A hammer stapler greatly speeds installation of building paper. Watch for any loose nails missed during the tear off and nail down any protruding staples. Avoid walking on building paper; it is slippery and can easily tear away from its staples. Some roofers opt to apply one course of building paper at a time, applying four or five courses of shingles before rolling out the next course of building paper.

Tools & Materials ▸

Chalkline	Caulk gun
Hammer stapler	30# building paper
Flat pry bar	Ice and water shield
Utility knife	Staples
Tape measure	Roofing cement

For optimum roof protection, apply ice and water shield in valleys, along the eaves, and along the rake edges of the roof. Apply 30# building paper over the remainder of the roof.

How to Install Underlayment

Snap a chalkline 35⅝" up from the eaves, so the first course of the 36"-wide membrane will overhang the eaves by ⅜". Install a course of ice and water shield, using the chalkline as a reference, and peeling back the protective backing as you unroll it.

Measuring up from the eaves, make a mark 32" above the top of the last row of underlayment, and snap another chalkline. Roll out the next course of building paper (or ice guard, if required) along the chalkline, overlapping the first course by 4". *Tip: Drive staples every 6 to 12" along the edges of building paper, and one staple per sq. ft. in the field area.*

At valleys, roll building paper across from both sides, overlapping the ends by 36". Install building paper up to the ridge—ruled side up—snapping horizontal lines every two or three rows to check alignment. Overlap horizontal seams by 4", vertical seams by 12", and hips and ridges by 6". Trim the courses flush with the rake edges.

Apply building paper up to an obstruction, then resume laying the course on the opposite side (make sure to maintain the line). Cut a patch that overlaps the felt paper by 12" on all sides. Make a crosshatch cutout for the obstruction. Position the patch over the obstruction, staple it in place, then trim away the crosshatch flaps for a close fit.

At the bottom of dormers and sidewalls, tuck the felt paper under the siding where it intersects with the roof. Carefully pry up the siding and tuck at least 2" of paper under it. Also tuck the paper under counter flashing or siding on chimneys and skylights. Leave the siding or counter flashing unfastened until after you install the step flashing.

Drip Edge

Drip edge is a flashing that's installed along the eaves and rake edges of the roof to direct water away from the roof decking. Although its job is to deflect water, it also gives the edges of the roof an attractive finish. A corrosion-resistant material, drip edge won't stain your roofing materials or fascia.

The flashing is installed along the eaves before the building paper is attached to allow water to run off the roof in the event it gets under the shingles. Drip edge is installed at the rake edges after the building paper has been attached to keep wind-driven rain from getting under the paper.

Drip edge is always nailed directly to the roof decking, rather than to the fascia or rake boards. The nail heads are later covered by roofing materials.

There are two basic styles of drip edge. One is the C-style drip edge that doesn't have an overhang, and the other, much more common, type is the extended-profile drip edge that has a hemmed overhang along the edges.

Tools & Materials ▸

Hammer	Roofing nails
Tape measure	Circular saw
Aviation snips	30# building paper
Drip edge	Ice and water shield

Drip edge

Drip edge flashing prevents water from working its way under the roofing materials along the eaves and rake edges of the roof.

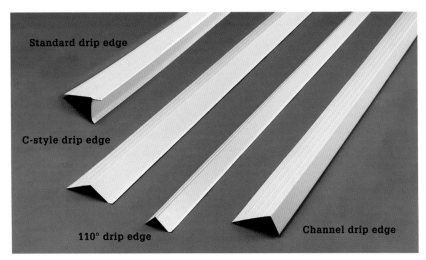

Standard drip edge

C-style drip edge

110° drip edge

Channel drip edge

Drip-edge molding is made of powder-coated aluminum, generally sold in 10-ft. lengths. Stock colors at most building centers are white and dark brown. It comes in several profiles, although only the standard and channel types (photo left) are stocked consistently. Take care not to confuse drip cap molding with edge-drip cap, a similar product, also sold in 10-ft. lengths, that is installed above windows and doors.

How to Install Drip Edge

Cut a 45° miter at one end of the drip edge using aviation snips. Place the drip edge along the eaves end of the roof, aligning the mitered end with the rake edge. Nail the drip edge in place every 12".

Overlap pieces of drip edge by 2". Install drip edge along the eaves, ending with a mitered cut at the opposite end.

Apply building paper, and ice guard if needed, to the roof, overhanging the eaves by ⅜" (see pages 44 to 45).

Cut a 45° miter in a piece of drip edge and install it along the rake edge, forming a miter joint with the drip edge along the eaves. Overlap pieces by 2", making sure the higher piece is on top at the overlap. Apply drip edge all the way to the peak. Install drip edge along the other rake edges the same way.

Flashing

Flashing is a metal or rubber barrier used to protect the seams around roof elements or between adjoining roof surfaces. Metal flashings are made of either galvanized steel, copper, or aluminum. Whatever metal you choose, use nails made of the same material. Mixing metals can cause corrosion and discoloration.

Flashing's primary job is to channel water off the roof and away from seams. It's installed in areas where shingles can't be applied and would otherwise be prone to leaks. Some flashing, such as the valley flashing shown on the opposite page, is installed over the underlayment, prior to the installation of the shingles. Other flashing, such as flashing for vent pipes, is installed in conjunction with the shingles, and is shown as part of the roofing sequences throughout this chapter.

While most flashing is preformed, you'll sometimes need to bend your own. This is especially true for flashing around roof elements, such as chimneys and dormers, that often need to be custom fit. Building a bending jig, as shown on the opposite page, allows you to easily bend flashing to fit your needs.

When installing flashing around roof elements, the flashing should be secured to one surface only—usually the roof deck. Use only roofing cement to bond the flashing to the roof elements. The flashing must be able to flex as the roof element and the roof deck expand and contract. If the flashing is fastened to both the roof deck and roof element, it will tear or loosen.

Flashing is a critical component of roofs that helps keep the structure watertight. Most roofs have flashing in the valleys and around dormers. This roof uses several valley flashings as well as flashing around the window and around the bump-out in the roof.

Metal roof flashings come in numerous profiles and shapes for specific purposes. Common flashings include: (A) roll flashing, also called handy flashing, made from galvanized metal; common widths range from 4" to 20"; (B) valley flashing (powder-coated with preformed W spine is shown); (C) drip cap molding installed above windows and doors; (D) flexible step molding for flashing around corners of vertical objects; (E) step flashing; (F) bent step flashing; (G) aluminum roll flashing; (H) copper roll flashing.

How to Bend Flashing

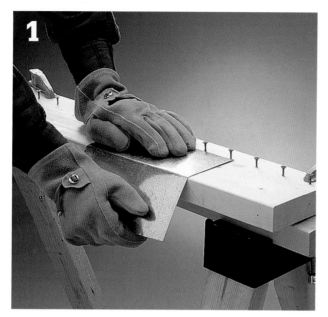

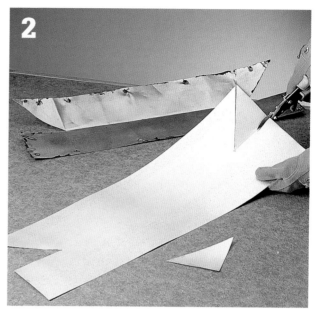

To bend flashing, make a bending jig by driving screws into a piece of wood, creating a space one-half the width of the flashing when measured from the edge of the board. Clamp the bending jig to a work surface. Lay a piece of flashing flat on the board and bend it over the edge.

Use the old flashing as a template for making replacement pieces. This is especially useful for reproducing complicated flashing, such as saddle flashing for chimneys and dormers.

How to Install Valley Flashing

8" overlap

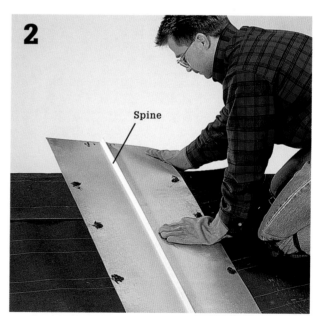

Spine

Starting at the eaves, set a piece of valley flashing into the valley so the bottom of the V rests in the crease of the valley. Nail the flashing at 12" intervals along each side. Trim the end of the flashing at the eaves so it's flush with the drip edge at each side. Working toward the top of the valley, add flashing pieces so they overlap by at least 8" until you reach the ridge.

Let the top edge of the flashing extend a few inches beyond the ridge. Bend the flashing over the ridge so it lies flat on the opposite side of the roof. If you're installing preformed flashing, make a small cut in the spine for easier bending. Cover nail heads with roofing cement (unless you're using rubber gasket nails). Apply roofing cement along the side edges of the flashing.

Asphalt Shingles

If you want to install asphalt shingles on your roof, then you're in good company. Asphalt shingles, also known as composition shingles, are the roofing of choice for nearly four out of five homeowners in America. They perform well in all types of climate, are available in a multitude of colors, shapes, and textures to complement every housing design. They also are less expensive than most other roofing products.

Asphalt shingles are available as either fiberglass shingles or organic shingles. Both types are made with asphalt, the difference being that one uses a fiberglass reinforcing mat, while the other uses a cellulose-fiber mat. Fiberglass shingles are lighter, thinner, and have a better fire rating. Organic shingles have a higher tear strength, are more flexible in cold climates, and are used more often in northern regions.

Although the roofing market has exploded with innovative new asphalt shingle designs, such as the architectural or laminated shingle that offers a three-dimensional look, the standard three-tab asphalt shingle is still the most common. The tabs provide an easy reference for aligning shingles for installation.

To help the job get done faster, rent an air compressor and pneumatic roofing gun. This will greatly reduce the time you spend nailing.

Tools & Materials ▸

Aviation snips	Tape measure
Carpenter's square	Caulk gun
Chalkline	Flashing
Flat pry bar	Shingles
Roofer's hammer or	Nail coils
pneumatic nailer	Roofing cement
Utility knife	Roofing nails
2-in-1 roofing knife	(⅞", 1¼")
Straightedge	Rubber gasket nails

Stagger shingles for effective protection against leaks. If the tab slots are aligned in successive rows, water forms channels, increasing erosion of the mineral surface of the shingles. Creating a 6" offset between rows of shingles—with the three-tab shingles shown above—ensures that the tab slots do not align.

How to Install Three-tab Shingles

Full tab

Half tab

Cover the roof with building paper (pages 44 to 45) and install drip edge (pages 46 to 47). Snap a chalkline onto the building paper or ice guard 11½" up from the eaves edge, to mark the alignment of the starter course. This will result in a ½" shingle overhang for standard shingles. *Tip: Use blue chalk rather than red. Red chalk will stain roofing materials.*

Trim off one-half of an end tab on a shingle. Position the shingle upside down, so the tabs are aligned with the chalkline and the half-tab is flush against the rake edge. Drive ⅞" roofing nails near each end, 1" down from each slot between tabs. Butt a full upside-down shingle next to the trimmed shingle, and nail it. Fill out the row, trimming the last shingle flush with the opposite rake edge.

First full course

Apply the first full course of shingles over the starter course with the tabs pointing down. Begin at the rake edge where you began the starter row. Place the first shingle so it overhangs the rake edge by ⅜" and the eaves edge by ½". Make sure the top of each shingle is flush with the top of the starter course, following the chalkline.

Snap a chalkline from the eaves edge to the ridge to create a vertical line to align the shingles. Choose an area with no obstructions, as close as possible to the center of the roof. The chalkline should pass through a slot or a shingle edge on the first full shingle course. Use a carpenter's square to establish a line perpendicular to the eave edge.

(continued)

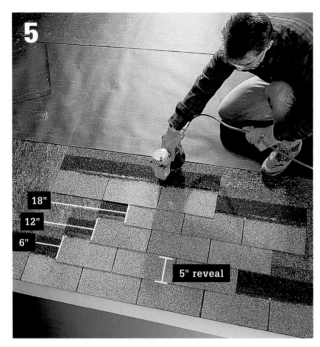

Use the vertical reference line to establish a shingle pattern with slots that are offset by 6" in succeeding courses. Tack down a shingle 6" to one side of the vertical line, 5" above the bottom edge of the first-course shingles to start the second row. Tack down shingles for the third and fourth courses, 12" and 18" from the vertical line. Butt the fifth course against the line.

Fill in shingles in the second through fifth courses, working upward from the second course and maintaining a consistent 5" reveal. Slide lower-course shingles under any upper-course shingles left partially nailed, and then nail them down. *Tip: Install roof jacks, if needed, to improve footing after filling out the fifth course.*

Check the alignment of the shingles after each four-course cycle. In several spots on the last installed course, measure from the bottom edge of a shingle to the nearest felt paper line. If you discover any misalignment, make minor adjustments over the next few rows until it's corrected.

When you reach obstructions, such as dormers, install a full course of shingles above them so you can retain your shingle offset pattern. On the unshingled side of the obstruction, snap another vertical reference line using the shingles above the obstruction as a guide.

Shingle upward from the eaves on the unshingled side of the obstruction using the vertical line as a reference for re-establishing your shingle slot offset pattern.

Trim off excess shingle material at the V in the valley flashing using a utility knife and straightedge. Do not cut into the flashing. The edges will be trimmed back farther at a slight taper after both roof decks are completely shingled.

Install shingles on adjoining roof decks, starting at the bottom edge, using the same offset alignment pattern shown in steps 1 to 6. Install shingles until courses overlap the center of the valley flashing. Trim shingles at both sides of the valley when finished. Trim shingles overhanging the rake, or trim each with a utility knife and a speed square as you install it.

Install shingles up to the vent pipe so the flashing rests on at least one row of shingles. Apply a heavy double bead of roofing cement along the edges of the flange.

(continued)

Place a flashing boot over the vent pipe. Position the flashing collar so the longer portion of the tapered neck slopes down the roof and the flange lies over the shingles at the base. Nail the perimeter of the flange using rubber gasket nails.

Cut shingles to fit around the neck of the flashing so they lie flat against the flange. Do not drive roofing nails through the flashing. Instead, apply roofing cement to the back of shingles where they lie over the flashing.

Shingle up to an element that requires flashing so the top of the reveal areas are within 5" of the element. Install base flashing using the old base flashing as a template. Bend a piece of step flashing in half and set it next to the lowest corner of the element. Mark a trim line on the flashing, following the vertical edge of the element. Cut the flashing to fit.

Pry out the lowest courses of siding and any trim at the base of the element. Insert spacers to prop the trim or siding away from the work area. Apply roofing cement to the base flashing in the area where the overlap with the step flashing will be formed. Tuck the trimmed piece of step flashing under the propped area, and secure the flashing. Fasten the flashing with one rubber gasket nail driven near the top and into the roof deck.

Apply roofing cement to the top side of the first piece of step flashing where it will be covered by the next shingle course. Install the shingle by pressing it firmly into the roofing cement. Do not nail through the flashing underneath.

Tuck another piece of flashing under the trim or siding, overlapping the first piece of flashing at least 2". Set the flashing into roofing cement applied on the top of the shingle. Nail the shingle in place without driving nails through the flashing. Install flashing up to the top of the element the same way. Trim the last piece of flashing to fit the top corner of the element. Reattach the siding and trim.

Counterflashing

Base flashing

Shingle up to the chimney base. Use the old base flashing as a template to cut new flashing. Bend up the counter flashing. Apply roofing cement to the base of the chimney and the shingles just below the base. Press the base flashing into the roofing cement and bend the flashing around the edges of the chimney. Drive rubber gasket nails through the flashing flange into the roof deck.

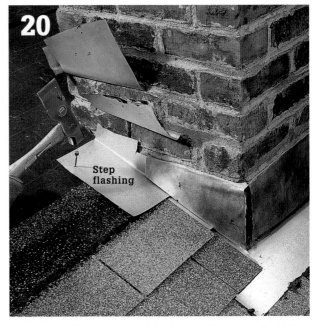

Step flashing

Install step flashing and shingles, working up to the high side of the chimney. Fasten flashing to the chimney with roofing cement. Fold down the counter flashing as you go.

(continued)

Cut and install top flashing (also called a saddle) around the high side of the chimney. Overlap the final piece of flashing along each side. Attach the flashing with roofing cement applied to the deck and chimney and with rubber gasket nails driven through the flashing base into the roof deck. Shingle past the chimney using roofing cement (not nails) to attach shingles over the flashing.

When you reach a hip or ridge, shingle up the first side until the top of the uppermost reveal area is within 5" of the hip or ridge. Trim the shingles along the peak. Install shingles on the opposite side of the hip or ridge. Overlap the peak no more than 5".

Cut three 12"-sq. cap shingles from each three-tab shingle. With the back surface facing up, cut the shingles at the tab lines. Trim the top corners of each square with an angled cut, starting just below the seal strip to avoid overlaps in the reveal area.

Snap a chalkline 6" down from the ridge, parallel to the peak. Attach cap shingles, starting at one end of the ridge, aligned with the chalkline. Drive two 1¼" roofing nails per cap about 1" from each edge, just below the seal strip.

Following the chalkline, install cap shingles halfway along the ridge, creating a 5" reveal for each cap. Then, starting at the opposite end, install caps over the other half of the ridge to meet the first run in the center. Cut a 5"-wide section from the reveal area of a shingle tab, and use it as a "closure cap" to cover the joint where the caps meet.

Shingle the hips in the same manner using a chalk reference line and cap shingles. Start at the bottom of each hip and work to the peak. Where hips join with roof ridges, install a custom shingle cut from the center of a cap shingle. Set the cap at the end of the ridge and bend the corners so they fit over the hips. Secure each corner with a roofing nail, and cover the nail heads with roofing cement.

After all shingles are installed, trim them at the valleys to create a gap that's 3" wide at the top and widens at a rate of 1/8" per foot as it moves downward. Use a utility knife and straightedge to cut the shingles, making sure not to cut through the valley flashing. At the valleys, seal the undersides and edges of shingles with roofing cement. Also cover exposed nail heads with roofing cement.

Mark and trim the shingles at the rake edges of the roof. Snap a chalkline 3/8" from the edge to make an overhang, then trim the shingles using a utility knife or an old pair of scissors.

Shingling Over an Old Roof

Installing shingles over your current shingles saves you the time, labor, and expense of tearing off the old roof covering. This method certainly has its appeal, but there's still some preparation work involved. Make any necessary repairs to the roof decking before applying new shingles. Replace any missing shingles, and renail any loose ones. Drive down protruding nails so the heads won't pierce the new roofing materials.

In order to shingle over old roofing, you cannot have more than one or two layers of shingles already on the roof, depending on your building codes. If you already have the maximum number of layers, the old shingles will need to be completely removed. To check for underlying shingles, lift up the shingles along the rake or eaves end of the roof and count the number of layers.

Before starting the project, read the section on shingling a roof, pages 50 to 57.

Before starting the project, read the section on shingling a roof, pages 50 to 57.

Tools & Materials ▸

Aviation snips
Carpenter's square
Chalkline
Flat pry bar
Roofer's hatchet or pneumatic nailer
Utility knife
Straightedge
Tape measure
Flashing
Shingles
Roofing cement
Roofing nails

Installing shingles over an old roof is frowned upon in some quarters, primarily because it does not allow you to inspect the roof deck and underlayment. But if your old roof is in good condition, with no more than two layers of shingles, most municipalities will allow you to add another layer.

How to Shingle Over an Old Roof

Cut tabs off shingles and install the remaining strips over the reveal area of the old first course, creating a flat surface for the starter row of new shingles. Use roofing nails that are long enough to penetrate the roof decking by at least ¾".

Trim the top of shingles for the first course. The shingles should be sized to butt against the bottom edge of the old third course, overhanging the roof edge by ½". Install shingles so the tab slots don't align with the slots in the old shingles.

Using the old shingles to direct your layout, begin installing the new shingles. Maintain a consistent tab/slot offset if you are installing three-tab shingles. Shingle up toward the roof ridge, stopping before the final course. Install flashing as you proceed.

Flashing is installed using the same techniques and materials as shingling over felt paper, except that you need to trim or fill in shingles around vent pipes and roof vents to create a flat surface for the base flange of the flashing pieces. *Tip: Valley flashing in good condition does not have to be replaced. Replace any other old flashing as you go.*

Tear off old hip and ridge caps before shingling these areas. Replace with new ridge caps after all other shingling has been completed.

Ridge Vents

For efficient attic ventilation, installing a continuous ridge vent is a reliable solution. Since they're installed along the entire ridge of the roof, they provide an even flow of air along the underside of the roof decking. Combined with continuous soffit vents, this is the most effective type of roof ventilation system.

Since the vents are installed along the ridge, they don't interrupt the field shingles, eliminating any disruptions to the roof. Other vent types, such as roof louvers and turbines, can distract from the roof's aesthetics.

Installing one continuous ridge vent is quicker and easier than installing other types of vents that need to be placed in several locations across the roof. It also saves you from having to make numerous cuts in your finished roof, which can disturb surrounding shingles.

Continuous ridge vents work in conjunction with the soffits to allow airflow under the roof decking. Installed at the roof peak and covered with cap shingles, ridge vents are less conspicuous than other roof vents.

Tools & Materials ▸

Hammer
Circular saw
Tape measure
Chalkline

Flat pry bar
Ridge vents
1½" roofing nails

How to Install a Ridge Vent

Remove the ridge caps using a flat pry bar. Measure down from the peak the width of the manufacturer's recommended opening, and mark each end of the roof. Snap a chalkline between the marks. Repeat for the other side of the peak. Remove any nails in your path.

Set the blade depth of a circular saw to cut the sheathing but not the rafters. Cut along each chalkline, staying 12" from the edges of the roof. Remove the cut sheathing using a pry bar. If you have a long ridge, spare your saw blade by first cutting away the roofing with a utility knife.

Measure down from the peak half the width of the ridge vent, and make a mark on both ends of the roof. Snap a line between the marks. Do this on both sides of the peak.

Center the ridge vent over the peak, aligning the edges with the chalklines. Install using roofing nails that are long enough to penetrate the roof sheathing. *Tip: If a chimney extends through the peak, leave 12" of sheathing around the chimney.*

Butt sections of ridge vents together and nail the ends. Install vents across the entire peak, including the 12" sections at each end of the roof that were not cut away.

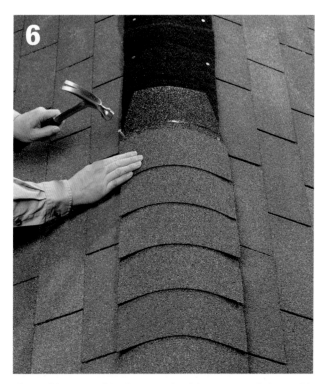

Place ridge cap shingles over the ridge vents. Nail them with two 1½" roofing nails per cap. Overlap the caps as you would on a normal ridge. If the caps you removed in step 1 are still in good shape, you can reuse them. Otherwise, use new ones.

Cedar Shakes

Cedar shakes (which are thick and rough) and shingles (which are tapered and smooth) are installed in much the same way, with one major difference. Shakes have felt paper installed between each course, while shingles do not. Shingles are often applied over open sheathing, while shakes are installed over open or solid sheathing. Air circulation under shakes and shingles can increase their life span. Check your local building codes to see what type of sheathing is recommended for your area.

The gaps between shakes and shingles, called joints, are specified by the manufacturer. You can determine how much of the material to leave exposed below the overlap, as long as it falls within the manufacturer's guidelines.

Tools & Materials ▸

Roofer's hatchet
Tape measure
Utility knife
Stapler
Chalkline
Circular saw
Jigsaw
Caulk gun
Shakes
Flashing
Nails

30# building paper
Stapler
Mason's string
Roofing cement

Cedar shakes are an attractive choice that has lately fallen out of favor because of high cost and maintenance challenges. Added to their premium price is a boost in insurance costs because in some regions they are considered a fire hazard.

Cedar Shakes & Shingles

Wood shakes and shingles are available in different grades. Some of the more popular include resawn shake (A), No. 1 hand-split medium shake (B), standard-grade shake (C), taper-sawn shake (D), No. 1 heavy shake (E), pressure-treated medium shake (F), No. 2 shingle (G), undercoursing shingle (H), No. 1 shingle (I).

Underlayment for Cedar Shakes & Shingles

Spaced sheathing is common, and sometimes required, for cedar shakes and shingles. Solid sheathing is installed along the eaves and rake ends, and open-spaced wood strips are installed in the field to allow for air circulation.

1 × 6 nailing strip

2 × 4 on flat

Shingle exposure

To install spacer strips over solid sheathing, place 2 × 4s flat over each rafter and nail them to the roof. Nail 1 × 4 or 1 × 6 nailing strips across the 2 × 4s. Keep the strips together along the eaves, then space them at a distance equal to the exposure rate in the field.

How to Install Cedar Shakes

Prepare the roof decking by installing valley flashing at all valleys (page 49). Apply building paper underlayment to the first 36" of the roof deck. *Note: Depending on your climate and building codes, you may need to install ice and water shield for this step rather than felt paper.*

Install a starter shake so it overhangs the eaves and rake edge by 1½". Do the same on the opposite side of the roof. Run a taut string between the bottom edges of the two shakes. Install the remaining shakes in the starter row, aligning the bottoms with the string. Keep the manufacturer's recommended distance between shakes, usually ⅜ to ⅝".

Set the first course of shakes over the starter row, aligning the shakes along the rake ends and bottoms. Joints between shakes must overlap by at least 1½". Drive two nails in each shake, ¾ to 1" from the edges, and 1½ to 2" above the exposure line. Use the hatchet to split shakes to fit. *Tip: Set the gauge on your roofer's hatchet to the exposure rate. You can then use the hatchet as a quick reference for checking the exposure.*

Snap a chalkline over the first course of shakes at the exposure line. Snap a second line at a distance that's twice the exposure rate. Staple an 18"-wide strip of felt paper at the second line. Overlap felt paper vertical seams by 4". Install the second course of shakes at the exposure line, offsetting joints by 1½" minimum. Install remaining courses the same way.

Set shakes in place along valleys, but don't nail them. Hold a 1 × 4 against the center of the valley flashing without nailing it. Place it over the shakes to use as a guide for marking the angle of the valley. Cut the shakes using a circular saw, then install.

Use the 1 × 4 to align the edge of the shakes along the valley. Keep the 1 × 4 butted against the valley center, and place the edge of the shake along the edge of the board. Avoid nailing through the valley flashing when installing the shakes.

Notch shakes to fit around a plumbing stack using a jigsaw, then install a course of shakes below the stack. Apply roofing cement to the underside of the stack flashing, then place it over the stack and over the shakes. Nail the flashing along the edges.

Overlap the exposed flashing with the next row of shakes. Cut notches in the shakes to fit around the stack, keeping a 1" gap between the stack and shakes.

(continued)

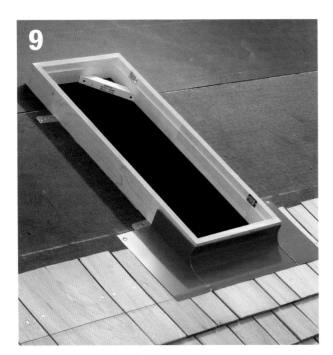

Install shakes under the bottom apron flashing beneath a skylight. Cut the shakes as necessary. Nail the shakes without driving nails through the flashing. Apply roofing cement to the underside of the flashing, then press to the shakes.

Flashing

Interweave skylight flashing along the skylight with rows of shakes. After each row of shakes, install a piece of flashing with the vertical plane placed under the edge lip of the skylight and the horizontal plane flush with the bottom edge of the shake. A row of shakes covers the top apron flashing.

Apply roofing cement along the underside of the roof louver flange, then set it over the vent cutout and over the shakes directly below it. Nail the louver in place. Install shakes over the sides and back of the louver, trimming to fit as needed.

As you approach the ridge, measure from the last installed row to the peak. Do this on each side of the roof. If the measurements are not equal, slightly adjust the exposure rate in successive rows until the measurements are the same. Make sure you're measuring to points that are aligned at the peak. The top of the sheathing is probably not level across the roof and cannot be a reference point.

Run shakes past the roof peak. Snap a chalkline across the shakes at the ridge. Set the circular saw blade to the depth of the shakes, then cut along the chalkline.

Cut 8" strips of felt paper and staple them over the hips and ridge. Set a factory-made hip and ridge cap at one end of the ridge, aligned with the roof peak. Do the same at the other end of the roof. Snap a chalkline between the outside edges of the caps.

Exposure rate

Set a ridge cap along the chalkline flush with the edge of the roof to serve as the starter. Install with two nails. Place a cap directly on top of the starter cap, and nail in place. Install caps along the remainder of the ridge, alternating the overlap pattern. The exposure rate should be the same as the roof shakes. Nails should penetrate the roof decking by ½".

Variation: If the ridge caps are not preassembled by the manufacturer, install the first cap along the chalkline, then place the second cap over the edge of the first. Alternate the overlap pattern across the ridge.

Roll Roofing

Roll roofing is a quick and easy roofing product to install. In the most simple installation, the material is rolled across the roof, nailed along the edges, and sealed with roofing cement. It's geared for roofs with slight slopes, such as porches and garages.

Some manufacturers recommend using a roof primer prior to installing the roofing. Read and follow manufacturer's directions. Your roof decking must be completely clean before the roll roofing can be applied. Any debris, even a small twig or leaf, can end up showing through the roofing.

Store the roofing in a warm, dry location until you're ready to start the project, and choose a warm day for the installation. Roll roofing is best installed in temperatures above 45°F. If applied in cold weather, the material can crack.

The following pages show the four methods for installing roll roofing. The perimeter bond application (pages 69 to 70) is the fastest installation method and can be used on sloped roofs. The concealed nail application (page 71) is best for roofs with a slighter pitch all the way down to a 1-in-12 slope, because it prevents water from penetrating under the nail heads. The double coverage method, also on page 71, is used for roofs that are almost completely flat. The double coverage, using fully bonded selvedge roofing, offers better protection against water infiltration. The last method is self-adhesive roll roofing (pages 72 to 73). Self-adhesive roll roofing is installed in much the same way as double coverage bonded roll roofing.

Tools & Materials ▸

Utility knife	Roll roofing
Tape measure	Galvanized
Chalkline	roofing nails
Serrated trowel	Asphalt-based
Straightedge	roofing cement
Hammer	Weighted roller

Roll roofing is used on roofs that have a slight slope. Installation is fast and straightforward, with the material rolled over a clean roof decking.

How to Install Roll Roofing in a Perimeter Bond

Nail drip edge along the eaves and rake ends of the roof (pages 72 to 73). Sweep the roof decking clean. Center an 18"-wide strip of roll roofing over the valley. Nail one side ¾" from the edge, every 6". Press the roofing firmly into the valley center, then nail the other side. Install a 36" strip over the valley the same way.

Snap a chalkline 35½" up from the eaves. Unroll the roofing along the chalkline, overhanging the eaves and rake edges by ½". Nail the roofing every 3" along the sides and bottom, ¾" from the edge of the decking. Roofing nails should be long enough to penetrate the roof decking by at least ¾".

Where more than one roll is needed to complete a course, apply roofing cement along the edge of the installed piece using a trowel. Place a new roll 6" over the first piece. Press the seam together and drive nails every 3" along the end lap. *Tip: Make sure the roofing is straight before nailing. Once it's nailed, you can't adjust it without creating wrinkles and folds. If it's running crooked, cut it and start with a new strip.*

Apply 2" of roofing cement along the top edge of the installed course. Install the second row flush with the line on the roofing, overlapping the cement edge. Drive nails every 3" along the rakes and overlap, ¾" from the edges. Do the same for remaining rows, offsetting seams at least 18".

(continued)

5

6

Cut roofing 1" from the valley center using a utility knife and straightedge. Be careful not to cut into the valley roofing. Apply a 6"-wide strip of roofing cement on the valley at the overlap. Place the main roofing over the cement. Do not nail closer than 12 inches from the center of the valley.

Install roofing in front of a vent pipe. Cut a square of roofing to fit over the pipe, with a hole in the center. Apply roofing cement around the edges of the square, then set it in place over the pipe. Overlap with the next row of roofing, notching for the pipe as necessary.

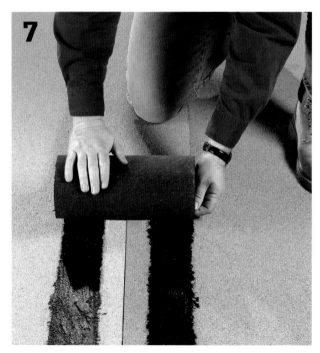

7

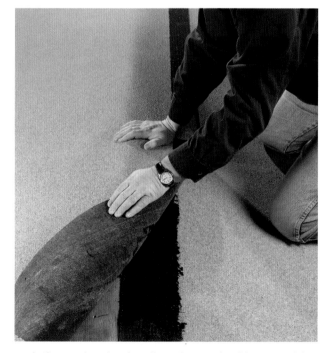

Cut the roofing flush with the roof peak. Snap a line on each side of the roof, 5½" from the peak. Apply 2" of cement above each line. Place a 12"-wide strip of roofing over the peak, flush with the chalklines. Drive nails every 3" along the seams, ¾" from the edges. *Tip: Use modest amounts of roofing cement. Excess cement can cause the roofing to blister.*

Variation: Rather than install a strip over the ridge, extend the roofing on one side of the roof 6" past the peak, overlapping the opposite side. Nail along the edge to secure it to the decking. Do the same on the other side, overlapping the installed roofing at the peak. Apply cement along the seam, and nail in place.

How to Install Roll Roofing (Concealed Nail Application)

Cut 9"-wide strips of roofing. Nail them in place along the rakes and eaves. Snap a chalkline 35½" up from the eaves. Place the first course of roofing flush with the line.

Nail the roofing along the top edge only, driving nails every 4", ¾" from the top edge. Roll back the side and bottom edges. Apply cement along the outside 2" of the strips installed in step 1. Set the overlapping strip back in place, pressing firmly to seal.

Set the next course in place so it overlaps the first row by 4". Nail along the top edge. Lift the side and bottom edges, apply cement, then press together to seal. Repeat this process for the remainder of the roof.

How to Install Roll Roofing (Double Coverage Application)

Cut away the granular part of the roofing to create a starter strip. Align the strip with the eaves, and drive nails along the top and bottom edge at 12" intervals. Place the first full course flush with the eaves. Nail the nongranule edges every 12". Roll back the bottom of the roofing and apply roofing cement along the eaves and rake edges on the starter strips. Set the roofing back in place, pressing it into the cement.

Align the bottom edge of the second course with the top of the granule edge of the first row. Nail every 12" along the nongranule edges. Flip the bottom part back, apply cement along the sides and bottom of the nongranule area of the first course, then set the strip back in place. Install remaining courses the same way.

How to Install Roll Roofing (Self-adhesive Application)

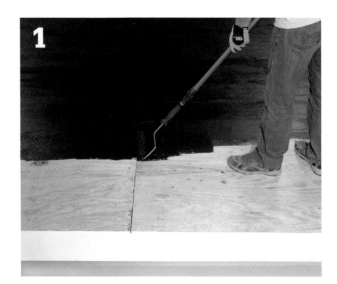

Although not necessarily required in all applications, the best adhesion is achieved when the roof is coated with primer, and some self-adhering roll roofing products require it anyway.

Cut a piece of base roll in half lengthwise about 19½" (about half of the meter-wide roll) and align it on the base of the roof, allowing about a 2" overhang or whatever is required to turn down and cover the fascia. When the base roll is centered and parallel to the roof, remove the backing and adhere it to the roof from the center of the roll, smoothing out any bubbles or wrinkling as you work toward the outside edges. Smooth the overhang onto the fascia.

For the next course, position the selvedge edge on the high side of the roof to provide an overlap guideline and to ensure water flows over the side laps. Fold the sheet in half away from the edge of the roof along its length, and remove the backing from the lower half of the sheet. From the center of the length, allow the sheet to roll onto the deck. Use a helper to maintain a straight line and sheet position, and hand-press the sheet to smooth out wrinkles and trapped air. Overlap subsequent sheets in the same course by at least 6", and cut any upper selvedge edges at 45° angles to avoid loose seams and to provide a smooth transition (inset). End laps in adjacent courses should be offset by at least 36".

Install subsequent courses by aligning the base ply with the guideline from the previous course. Fold the lower half of the sheet back onto itself once more, remove the backing, and work from the center, allowing the sheet to roll onto the deck. Smooth from the center outward. Repeat the process with the top half of the sheet. When the roof is covered, roll the surface with a weighted roller or push-broom. Use a helper to assist getting the roller or broom to the roof.

To install drip edge, coat the base with a thin layer of roofing cement, and set the drip edge into it. To reinforce the edge to the base ply, install roofing nails 3" on center along a staggered course through the drip edge. Use a notched trowel to apply a ⅛" to ¹⁄₁₆" layer of roofing cement to the top of the metal prior to installing the cap sheet. Also apply a layer of primer and flashing cement where flanged metal termination and penetration flashings are needed, as well as a layer of roofing cement to the top of any metal flanges prior to installing the cap sheets.

To apply the cap sheet, roll it out to manageable lengths, remembering that minimal seams is the goal. Allow the sheets to lay flat for 30 minutes, which lets the ends relax. On the low point of the roof, align the sheet to lay flat and parallel atop the drip edge and eaves, and in the same manner as the base sheet, fold the lower half of the sheet away from the edge of the roof, remove the backing, and allow it to roll onto the roof. Repeat with the upper half of the sheet.

To avoid blistering, apply a bead of roofing cement adhesive along the top edge of each cap sheet, and at any selvedge edge T-joints. On roofs with slopes greater than 1" in 12", backnail in the selvedge edge area at 18" on center with roof nails or cap nails. When additional lengths of cap sheet are needed in the same course, allow a minimum of 6" of overlap, and end-laps must be offset at least 36".

Complete overlapping end joints by applying a ⅛" to ¹⁄₁₆" layer of roofing cement to the granular surface of the underside of the sheet using a notched trowel. Flashing of walls, roof terminations, and obstructions can also be accomplished with these materials. Nail off the flashing at the top and counter-flash backwards. When the roof is completely covered, use a weighted roller to smooth out any wrinkles or bubbling. See photo, page 68.

EPDM Rubber Roofing

For roof decks with minimal pitch, even rolled asphalt roofing may not offer enough protection against leaks. In these situations, ethylene propylene diene monomer (EPDM) rubber membrane roofing may be your best—or only—bet. EPDM roofing is easy to install with minimal tools by a do-it-yourselfer. Unlike other membrane systems that must be applied with a torch, EPDM adheres with liquid adhesive. It comes in 10 × 20- or 20 × 100-ft. rolls so you can plan your installation to minimize seams.

Installing the membrane involves removing the previous roofing material down to bare roof decking and any flashings around vent pipes or other protrusions. Make sure your roof deck material is clean, dry, and in good repair. You may be able to overlay the deck with a layer of high-density fiberboard or 1" isocyanurate insulation board to create a fresh, flat deck surface if the previous surface isn't sufficiently flat or shows signs of minor deterioration. Avoid using insulation products with a waterproofing layer or film. Glue will not penetrate properly into the insulation.

Once the deck is prepared, lay out the membrane sheets so they overlap the edges of the roof and one another by 3". Make any necessary cutouts to allow for roof protrusions and to allow the membrane sheets to lay flat and relax. If your roof abuts a vertical wall, the membrane should extend up the wall 12" so it can be adhered to the wall and sealed with a metal termination bar.

Adhering the membrane to the roof deck involves applying liquid bonding adhesive onto the roof deck with the membrane rolled back, allowing it to partially cure, then pressing the membrane into place over the adhesive, and brushing it thoroughly to remove any air pockets. Once the membrane is fixed in place, you seal the overlapping seams with strips of soft seaming tape and liquid primer, rolling the seams flat. Finish up the roof by trimming the membrane at the roof edges, installing the appropriate boot flashings, and adding any termination bars that may be required.

Follow the EPDM manufacturer's instructions carefully, particularly if they differ from the step-by-step process you see here.

Tools & Materials ▶

Measuring tape	Liquid adhesive
Paint roller	Seaming tape
Stiff-bristle push broom	Primer
	Contact cement
Utility knife	Termination bars (if applicable)
J-roller or roller seaming tool	Exterior screws
EPDM membrane	

EPDM roofing provides the best protection against leaks on low-pitched roofs. It's an easy-to-install, DIY project. Most home centers now carry rubber roofing in standard 10 × 20 ft. sheets for less than $150. Some also will sell 10-ft. wide roofing by the linear foot from a longer roll. When possible, buy a large enough sheet to cover the entire roof. This greatly decreases the likelihood of leaks forming because it eliminates the need to seam the roof covering.

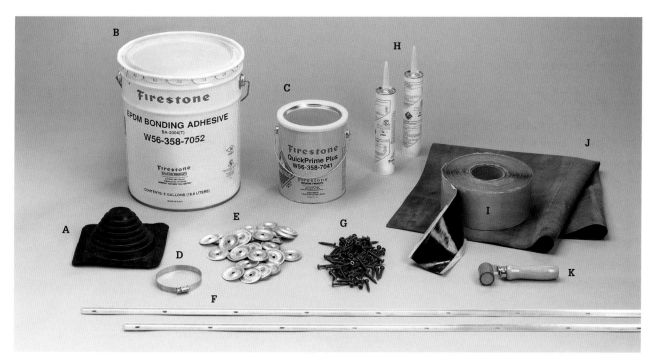

Rubber roof coverings require special adhesives and primers, as well as flashing and accessories such as rubber pipe boots. Shown here, from left to right, are pipe boots (A), EPDM adhesive and primer (B, C), hose clamp (D), insulation plates (E), termination bars (F), exterior screws (G), caulk (H), seaming tape (I), EPDM membrane (J), and J-roller (K).

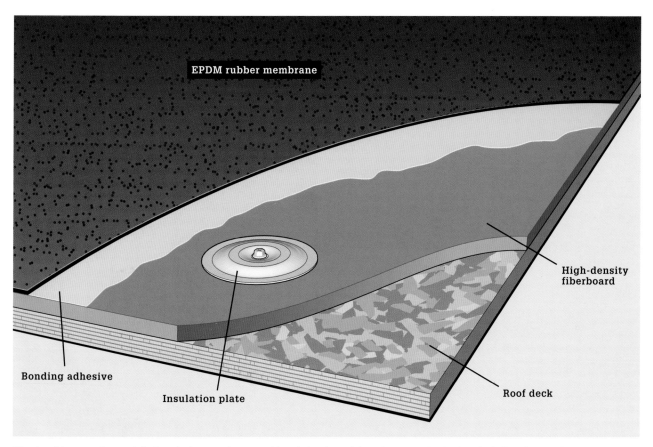

Most rubber roofs are fully bonded to a substrate of insulation board, but they can also be bonded directly to a plywood roof deck or even installed with a perimeter bond only. In some commercial applications they are installed with a layer of river rock on top for ballast.

How to Install an EPDM Rubber Roof

Prepare the roof deck for membrane roofing by removing the old roofing material down to bare decking. Look closely for signs of deterioration. Replace deteriorated decking. For a smooth surface, cover the roof with a new layer of high-density fiberboard (sold at roofing materials suppliers). Secure with fasteners recommended by the manufacturer for this purpose—usually long screws with large insulation plates.

Sweep the roof deck thoroughly, and spread out the membrane so it has a chance to relax. Make any cutouts in the membrane that may be necessary to allow for vent pipes or other protrusions. Overlap the sheets by 3" and wipe them down with the recommended cleaner to prepare the surfaces for adhesive.

Apply the latex adhesive. Fold half of the first membrane sheet over on itself to expose the roof deck, and roll a heavy coat of adhesive onto both the deck and membrane surfaces with a medium-nap paint roller. However, do not apply adhesive to the overlapped section of membrane. Once the adhesive begins to set (about 20 minutes in normal conditions), carefully roll the folded rubber down into place. Avoid wrinkling the membrane.

Use a stiff-bristle push broom to brush out any air pockets that may be evident under the bonded half of the membrane. Brush from the middle of the roof outward to the edges. Then fold the other unbonded half over, apply adhesive to the rubber and roof deck again, and adhere this half of the membrane to the roof. Apply all sections of membrane to the roof deck in this fashion, but do not apply adhesive within 3" of the edges of any overlapping sections of rubber; these must be accessible for applying seaming tape along the seams.

5

Roll the top section of overlapping membrane back along the seam area, and chalk a reference line 3" from the edge of the bottom membrane. This marks the area for applying seaming tape.

6

Tape the seams. Use the recommended cleaning solvent to clean both halves of the overlapping membrane in the tape areas, then apply seaming tape sticky side down to the bottom membrane within the marked area. Press the tape down firmly to ensure good adhesion to the membrane.

7

Fold the top membrane overlap back in place on the tape. Slowly pull off the tape's paper backing with the membrane edges now overlapping. Press the overlapping edges down to create a tighter, smooth seam. Roll the seamed areas with a J-roller or seam rolling tool to bond the seam (inset).

8

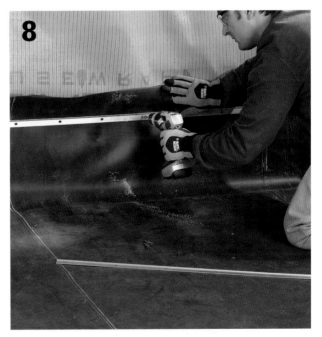

If the roof meets a vertical wall, you may need to remove siding so you can bond the membrane to wall sheathing. Use contact adhesive to apply the membrane 12" up the wall. Seal the edge with a metal termination bar fastened to the wall with exterior screws. Trim off overhanging membrane around the roof edges, and flash it according to the manufacturer's recommendations using rubber adhesive flashing and rubber boots.

Raised Seam Metal Roofing

Raised seam metal roofing is an attractive, long-lasting material that installs quickly and suits even the roughest climates. In addition to the standing seam metal roofing shown in the following steps, metal shingle and metal tile roofing materials are also available, and manufacturers offer a wide variety of colors for all three options.

Metal roofs are designed for fail-safe installation, as long as you follow the manufacturer's instructions and use the proper components. In fact, you are less likely to make a mistake installing metal roofing than you are with asphalt shingles, especially with valley and flashing installation. Some metal-working skills may be called for, but they are simple to learn.

If local codes permit and your old roof is smooth and in good condition, you can cover it with metal roofing without a tearoff. However, you must first check for structural damage, remove all moss and debris, remove protruding nails, and nail down loose or cupped roofing. In addition, remove all ridges, hips, and flashing, and cut back the roofing at eaves and rakes until it is flush with fascia.

New types of standing seam roofing have snaplock panels that suit slopes as low as 3-in-12. Others, whose seams are fastened on site using a simple tool, handle pitches as low as ½"-in-12.

Tools & Materials ▶

Cordless drill/driver
    with hex and Phillips bits
Snips, left-hand
Snips, right-hand
Snips, aviation type
Nibbler

Chalkline
Caulk gun
Pop riveter
Broad pliers (hand seamers)
Turn-up tool

Markers
Scratch awl
Utility knife
Mastic tape
Tape measure

Duckbill pliers
Hammer
Roofing nails
Eye protection
Gloves

Metal roofing is an attractive and long-lasting material. It installs easily and requires little maintenance.

How to Install Raised Seam Metal Roofing

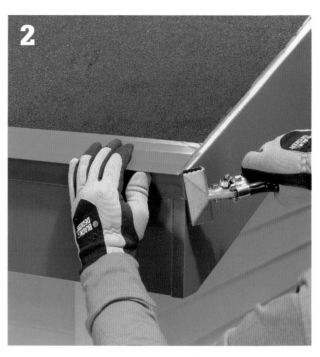

Cut and fasten drip edge after you have installed the ice and water shield and felt for the roof. Follow manufacturer's specs for the location of the fasteners, typically a few inches up from the edge of the eave. At a valley, cut the drip edge so it overlaps the adjacent piece.

Overlap drip edge at the gable so when the rake flashing is installed it also overlaps. To do this, cut away the lip of the flashing and score the bend line with an awl. Use duckbill pliers to grip the flashing at the score line and bend to make an overlapping tab. Use a pop riveter to secure the overlap.

Prepare the valley flashing by laying it on the valley and using a straight edge to mark a cut line along the eaves. Cut the flashing to create tabs that, once scored and bent, overlap the eaves flashing.

Position the valley flashing and push down so it is fully seated in the valley. Check that its end extends past the eaves by 1". Along the edge of each side, tack or screw it in place every 6".

Apply two rows of butyl mastic tape to each side of the valley ridge, spacing them according to the manufacturer's recommendations. Typically, the roofing panels overlap the flashing within a minimum of 4" of the ridge that runs down the middle of the valley flashing.

Use a helper to install the panels. The panels are durable once installed, but can be bent and scratched during handling. While the panels are waiting to be installed, weight them down or secure them on the ground to keep them from blowing away.

Not all roofs are square, so use a panel and tape measure to mark an alignment line ¼" from the eave. Use the "3-4-5 method" to check for square: Mark a point 3 ft. up from the eaves and another point 4 ft. from the panel. When the panel is squared up to the eaves, the distance between the two points is exactly 5 ft.

(continued)

8

Use a cordless drill driver with a socket to attach panels. Set the panel so it overhangs the eaves as recommended. Follow instructions for spacing the fasteners—every 12" on center is typical. Fasteners have flexible washers that must be snugged down firmly, but not too much.

9

Snap in the second panel. Metal panels install quickly, neatly locking in place with firm downward pressure. Begin locking the panel in place at the eaves and work upward. Never lock from both ends toward the center—you may create a bulge.

10

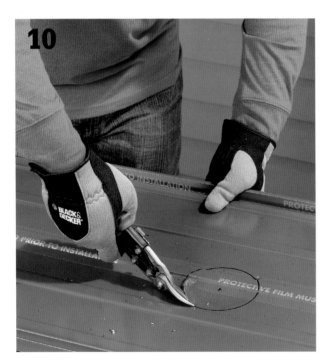

Cut openings for vent stacks and other obstacles with a metal snips or a nibbler. To mark the panel, set it beside the stack, check that it extends over the eaves by 1", and use a square to mark the upper and lower edges of the stack, adding ½" for wiggle room. Then measure from the nearest fastened panel so you can mark for the width of the stack.

11

Trim the vent stack boot so it fits snug over the stack. Neoprene boots are made with ridges that guide your cut. Dry fit the boot, adjust as needed, then refit it and caulk around the stack.

12

Caulk and fasten the base of the stack boot, positioning fasteners every 1–2". Double check that the caulk and boot seal along ridges in the panel.

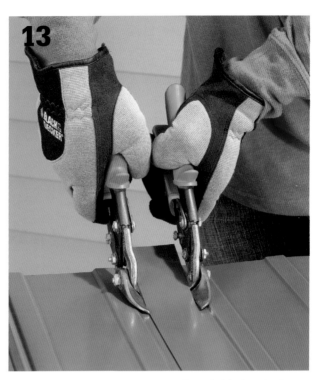

13

Cut panels lengthwise by using left- and right-hand snips in tandem. If you use a circular saw with a metal-cutting blade, add masking tape to the underside of the saw shoe to preventing scratching. Wear ear and eye protection.

14

Use a guide to mark panels at the valley. First, rough cut and position panels so they overhang the eaves properly. Then use a scrap of wood at least 4" wide as a marking guide. Cut the panel.

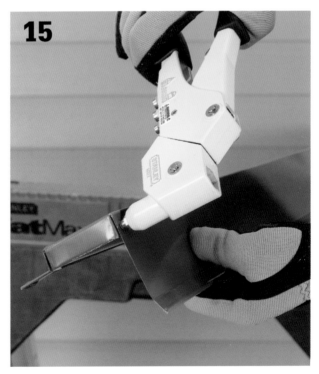

15

Close off the end of your rake flashing. Before cutting the flashing to length, cut tabs as shown and bend them so they overlap. Use pop rivets to fasten the tabs together. Trim the completed closure as needed.

(continued)

16

Fasten the rake flashing using the fastener and spacing recommended by the manufacturer—typically gasketed fasteners 24" on center (left). Apply butyl mastic tape before positioning and fastening.

Install side wall flashing after cutting tabs and bending them to close off the end. Install any end wall flashing as well. In both cases, you may first need to remove a course or two of siding in order to secure the flashing to the wall.

17

Install perforated zee closure top caps to the ridge to permit ventilation while preventing weather and pests from getting in. Check manufacturer's specs for positioning and caulking before fastening.

Close off the end of the ridge piece by cutting and bending tabs. Caulk the overlap before pop riveting the closures. Install the piece with fasteners driven into the standing seams.

Metal roofing panels snap in place one after another for a smooth, seamless look. *Note: Some manufacturers recommend leaving the top cap valley flashing off the finished roof in rainy or snowy areas.*

Faux slate looks just like the real thing but is easy to install and substantially less costly than slate. Made of polymer or recycled tires, the shingles are molded to mimic the rough edges and deep shadowlines of natural slate.

Faux Slate

No roofing materials are fun to work with, but faux (also called artificial or simulated) slate comes close. It is easy on the hands, cuts with a utility knife (though a power saw is faster), and nails easily. The shingles feel reassuringly hefty and firm, a material you can well imagine living up to its 50- to 75-year warranty. It does not soften in hot weather, easing summertime installation—a serious problem with asphalt shingles. Unlike the real thing, faux slate is an ideal do-it-yourself material as long as you abide by the manufacturer's installation instructions.

Although it is roughly twice as expensive as asphalt shingles, faux slate's good looks and longevity make it an increasingly popular choice, especially for owners of vintage homes. Some manufacturers tout the use of recycled materials; others emphasize "virgin" content unsullied by variable quality of recycled raw materials. Some faux slate shingles feel stiff and very plastic-like, while others are noticeably more flexible like rubber. Whatever the material, faux slate looks very much like the real thing.

Tools & Materials ▸

Tile bundle(s)
Chalkline
Extension ladder
Flashing
30# building paper
Drip cap
Ice and
    water shield
Pneumatic nailer
Galvanized
    roofing nails

Utility knife
Straightedge
Circular saw
Carbide saw blade
Sawhorse
    or workbench
Clamps
Fall-arresting gear
Eye protection
Gloves

Mix and Match to Suit your Roof ▸

Not all roofs are created equal, nor will match exactly the amount of tiles ordered. DaVinci, as we've used here, and other brands often are available in premixed and blended bundles of several different sizes. An installation where cuts don't need to be made along rakes or gable ends creates a more authentic-looking roof. When your measurement comes within 18 to 24" from a rake, you can get an exact measurement and install a combination of tiles that will fit without cutting. You'll reduce your waste by an incredible amount and also allow the realistic look of the faux slate to remain intact.

Faux tiles are available in premixed shapes and sizes to accommodate most roofs and reduce waste.

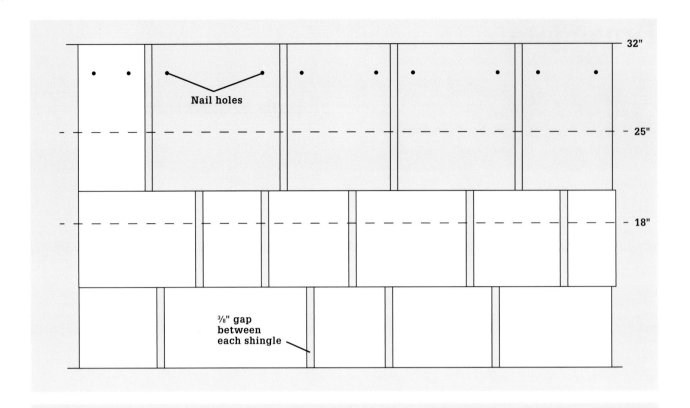

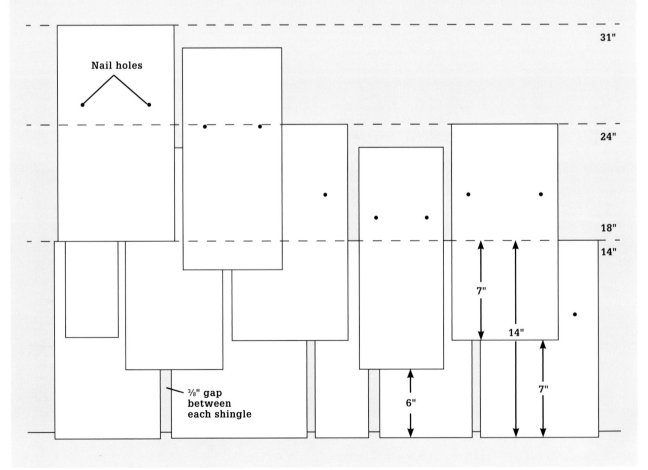

Faux slate shingles can be installed in straight coursing fashion or with staggered coursing.

How to Install Faux Slate Roofing

Prepare the roof by installing ice and water shield along the eaves and valleys. Cover the rest of the roof with 30# building paper and install drip cap. With a chalkline, strike a mark so your starter shingles will overhang the drip cap by 1".

Install the 12"-wide starter shingles, overhanging the eaves drip cap by 1". Nail the shingles where indicated, typically 6" up from the bottom and ¾" in from each edge. Use stainless steel, copper, or hot-dipped zinc 1½" roofing nails. Do not staple. Space shingles ⅜" apart, using a scrap of wood as a guide.

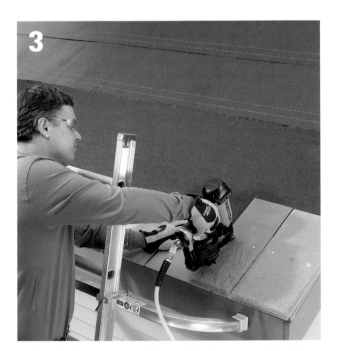

Begin the first course with any shingle except a 12"-wide shingle. This ensures that the gap between the starter shingles is overlapped 1½" or more. Nail the shingles where indicated on the surface.

Use the exposure alignment guides and nailing guides as you begin the second course. Always check that ⅜" gaps are maintained and overlapped by at least 1½" by shingles.

(continued)

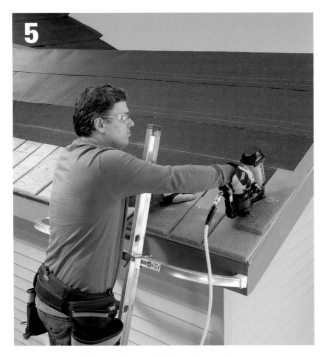

As a quick guide, strike a chalkline for each course. It is easier to spot while shingling, although it's worthwhile to confirm you are on track by occasionally checking the alignment guides on the faux shingles.

Try to avoid cutting a shingle at the end of a course. Instead, vary the gap from ³⁄₁₆" to ½" as you near the gable rake to finish a course without cutting. If a cut is necessary, use a utility knife and a straight edge or a circular saw equipped with a carbide-tipped blade, two teeth per inch.

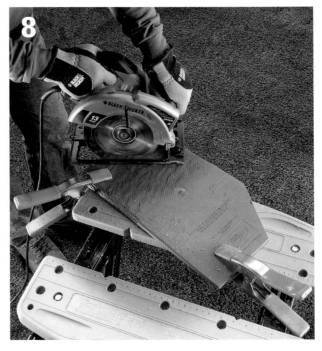

To make a closed valley, install 28"-wide "W" flashing. Run the courses to within a couple of shingles of the valley. Carefully measure so that a wide (10- or 12"-) shingle will be trimmed for the valley. Mark a cut line just shy of the base of the valley ridge. For an open valley, use "Double W" flashing (see page 48 for different types of flashing).

Cut the shingles where marked. Also make a 45° cut at the upper corner nearest the valley center. This helps divert toward the valley any water forced under the shingle.

Nail valley shingles, maintaining the ⅜" gap from the next shingle in the course. Nail no closer than 5" from the valley flashing ridge.

Cut starter triangles for hip ridges. After applying metal flashing to the ridge, apply the starter triangles. Nail 7" up from the bottom edge of the triangle and close to the outer edge of the flashing.

Use 7" shingles to complete the hip, fastening with two nails per shingle. Begin by lining up the shingle edge with the eave. Use a 6" exposure. Some manufacturers offer preformed hip and ridge shingles.

Shingle the ridge using 7" slates over continuous ridge vent using a 6" exposure. Vary the exposure slightly if needed to terminate the ridge with a shingle cut to 6".

Tile Roofing

Modern clay tiles use an S-type design rather than the two-piece system that was once common. This simplifies the installation process and saves you time. Due to the contour of the tiles, you'll need plumbing vents and air vents that match the shape of the roofing materials.

Before starting the project, make sure your roof framing can support the weight of the tiles. The materials are very heavy, and roofs designed for asphalt shingles may not have the structural support for clay tile. Check with your building inspector if you're unsure.

Tools & Materials ▸

Hammer	Sand
Tape measure	Portland cement
Chalkline	Lime
Circular saw	Premixed cement
Jigsaw	mortar or Type M
Trowel	Tile
Diamond saw blade	Bird stops
Caulk gun	Plumbing vents
30# felt paper	Air vents
Ice and water shield	Roofing sealant
Nailers (2 × 6,	Peel-and-stick
2 × 3, 2 × 2)	flashing
¾" roofing nails	Plastic cement
Roofing nails	

Tips for Installing a Tile Roof ▸

To cut clay tile, use a diamond blade in a circular saw or grinder. Clamp the tile to a work surface, make your cutting line on the tile, then cut along the line. Be sure to wear safety glasses and a respirator when making the cuts.

Mortar is available premixed, or you can mix it yourself. This project requires cement mortar Type M. To mix it, combine 3 parts portland cement, 1 part lime, and 2 parts sand. Add water and mix until the consistency is like mashed potatoes.

Clay tiles give homes a truly impressive roof that can't be imitated by other materials. The S design makes installation easier and less time-consuming. Each tile simply overlaps the preceding tile.

Clay tiles are beautiful, but be sure the roof framing is strong enough to support this heavier roofing material.

How to Install a Tile Roof

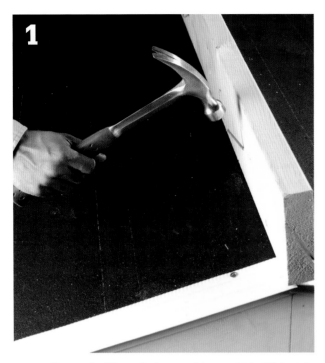

Cover the roof with underlayment (pages 44 to 45). Install drip edge (pages 46 to 47) and valley flashing (page 48). Nail 2 × 6 lumber on the edge over all ridges and hips.

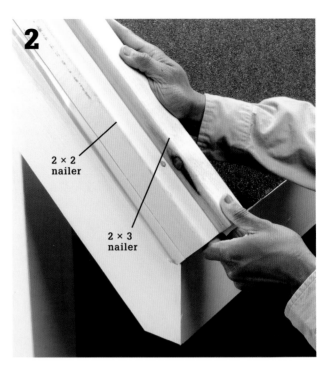

Install 2 × 2 nailers along the rake edges of the roof. Butt 2 × 3 nailers against the 2 × 2s and nail them in place.

2 × 2 nailer

2 × 3 nailer

Measuring from the outside edge of the 2 × 3 nailer along the left rake edge, make marks on the roof every 12". Center a bird stop over each mark, aligned with the front edge of the roof, and nail in place. *Note: Bird stops are available from the tile manufacturer, or you can cut your own from wood.*

Place gable tile over the 2 × 2s along the rake ends of the roof, overhanging the front of the roof by 3". Nail in place, using two ¾" roofing nails per tile. Overlap tiles by 3". *Note: Be sure to use left gable tiles for the left side and right gable tiles for the right side.*

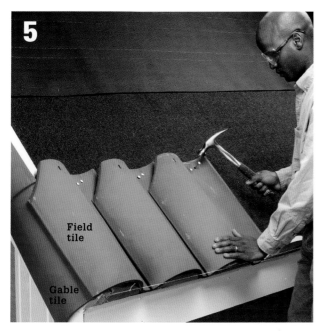

Starting on the left side of the roof, place the first field tile over the gable tile and 2 × 3 nailer. Align the end of the field tile with the end of the gable tile. Nail in place with two nails. Install the first course of tiles the same way, placing them over the bird stops. *Tip: To ensure alignment, tie a string across the end of the first gable tiles. Set the tiles flush with the string. Move the string to align subsequent rows.*

Install the next row of tiles on the roof, overlapping the first course by 3". Install remaining courses the same way. Avoid stepping on or walking on tiles as they break easily. When you can no longer reach new tiles from below, begin to work from higher up the roof. Work around obstacles as you encounter them, as seen in the remaining photos.

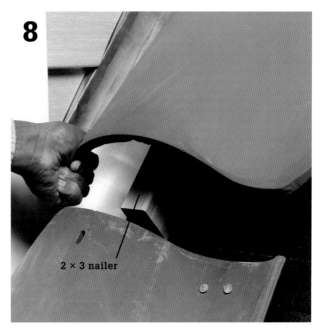

At dormers, chimneys, and walls, install pan flashing at least 4" up the wall and a minimum of 6" along the roof. Turn up the outside edge of the flashing 1½". Install counter flashing over the pan flashing. *Note: The top edge of the counter flashing must be installed under the wall siding or placed in the mortar between bricks in the chimney. The flashing may still be present from the old roof.*

Install a 2 × 3 nailer along the turned-up edge of the pan flashing. Set the tile over the nailer and nail it in place.

(continued)

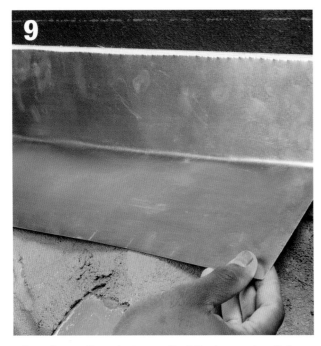

9

When the roofing abuts a wall of the house, install tile up the roof to the wall. Apply mortar generously between the tops of the tiles and the wall, filling in any gaps. Place 3 × 4 flashing over the mortar, then place counter flashing over that.

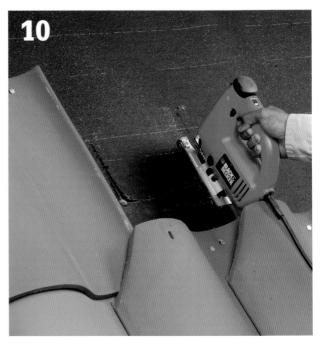

10

Mark roof vent locations between rows of tiles and between rafters. Follow manufacturer's guidelines for the size of the opening. Cut out the opening using a jigsaw or circular saw. *Tip: Periodically look at the roof from the street to make sure the rows are running straight and the tiles look uniform.*

11

Apply roofing sealant along the bottom of the primary roof vent, then install it over the opening using roofing nails driven every 4" through the flange. Seal the flange with peel-and-stick flashing. Place the secondary vent over the primary vent and nail it to the roof. Overlap the lip with the next piece of tile.

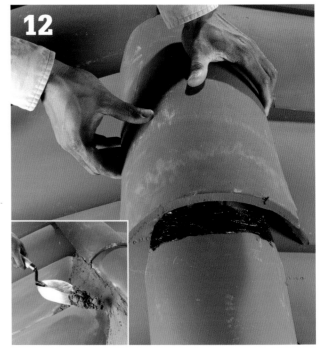

12

Center ridge tiles over the hips and peak. Apply a small amount of plastic cement on the nose of each ridge tile. Overlap the tiles for a 16" exposure, placing the tile over the plastic cement on the previous tile. Nail the ridge tiles using two nails per tile. Fill the gaps beneath ridge tiles with mortar (inset photo).

Tiling Around a Vent ▸

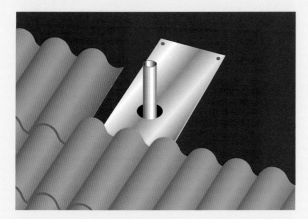

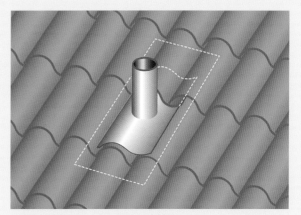

Place primary metal flashing over a plumbing vent and nail in the corners. Install tile over the front of the flashing.

Cover the back of the flashing with building paper. Install tile over the flashing and vent. Apply sealant along the underside of the secondary flashing, then place it over the tiles and vent. Overlap the back edge of the flashing with the next row of tiles.

Tile roofing is more expensive and labor-intensive to install, but few roofing materials offer a more stunning way to cover and highlight your house.

Inspecting & Repairing a Roof

A roof system is composed of several elements that work together to provide three basic, essential functions for your home: shelter, drainage, and ventilation. The roof covering and flashing are designed to shed water, directing it to gutters and downspouts. Air intake and outtake vents keep fresh air circulating below the roof sheathing, preventing moisture and heat buildup.

When your roof system develops problems that compromise its ability to protect your home—cracked shingles, incomplete ventilation, or damaged flashing—the damage quickly spreads to other parts of your house. Routine inspections are the best way to make sure the roof continues to do its job effectively.

Tools & Materials ▶

Tape measure
Wire brush
Aviation snips
Trowel
Flat pry bar
Hammer
Utility knife
Caulk gun
Replacement
   flashing

Replacement
   shingles
Roofing cement
Roofing nails
Plywood
Double-headed nails
Rubber gasket nails

Tips for Identifying Roofing Problems

Ice dams occur when melting snow refreezes near the eaves, causing ice to back up under the shingles, where it melts onto the sheathing and seeps into the house.

Inspect both the interior and the exterior of the roof to spot problems. From inside the attic, check the rafters and sheathing for signs of water damage. Symptoms will appear in the form of streaking or discoloration. A moist or wet area also signals water damage.

Common Roofing Problems

Wind, weather, and flying debris can damage shingles. The areas along valleys and ridges tend to take the most weather-related abuse. Torn, loose, or cracked shingles are common in these areas.

Buckled and cupped shingles are usually caused by moisture beneath the shingles. Loosened areas create an entry point for moisture and leave shingles vulnerable to wind damage.

A sagging ridge might be caused by the weight of too many roofing layers. It might also be the result of a more significant problem, such as a rotting ridge board or insufficient support for the ridge board.

Dirt and debris attract moisture and decay, which shorten a roof's life. To protect shingles, carefully wash the roof once a year using a pressure washer. Pay particular attention to areas where moss and mildew may accumulate. Seasonal application of moss killer is even more effective. Use only powders intended for roofs; grass types will stain roofing.

In damp climates, it's a good idea to nail a zinc strip along the center ridge of a roof, under the ridge caps. Minute quantities of zinc wash down the roof each time it rains, killing moss and mildew.

Overhanging tree limbs drop debris and provide shade that encourages moss and mildew. To reduce chances of decay, trim any limbs that overhang the roof.

How to Locate & Evaluate Leaks

If you have an unfinished attic, examine the underside of your roof with a flashlight on a rainy day. If you find wetness, discoloration, or other signs of moisture, trace the trail up to where the water is making its entrance.

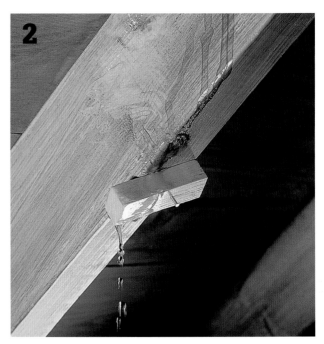

Water that flows toward a wall can be temporarily diverted to minimize damage. Nail a small block of wood in the path of the water, and place a bucket underneath to catch the drip. On a dry day, drive a nail through the underside of the roof decking to mark the hole.

If the leak is finding its way to a finished ceiling, take steps to minimize damage until the leak can be repaired. As soon as possible, reduce the accumulation of water behind a ceiling by poking a small hole in the wallboard or plaster and draining the water.

Once you mark the source of a leak from inside, measure from that spot to a point that will be visible and identifiable from outside the house, such as a chimney, vent pipe, or the peak of the roof. Get up on the roof and use that measurement to locate the leak.

How to Make Emergency Repairs

If your roof is severely damaged, the primary goal is to prevent additional damage until permanent repairs are made. Nail a sheet of plywood to the roof to serve as emergency cover to keep out the wind and water. *Tip: For temporary repairs, use double-headed nails, which can be easily removed. Fill nail holes with roofing cement when the repair is complete.*

Cover the damaged area by nailing strips of lath around the edges of a plastic sheet or tarp.

How to Make Spot Repairs with Roofing Cement

To reattach a loose shingle, wipe down the felt paper and the underside of the shingle. Let each dry, then apply a liberal coat of roofing cement. Press the shingle down to seat it in the bed of cement. *Tip: Heat softens the roof's surface, and cold makes it brittle. If needed, warm shingles slightly with a hair dryer to make them easier to work with and less likely to crack.*

Glue down buckled shingles by cleaning below the buckled area, filling the area with roofing cement, then pressing the shingle into the cement. Patch cracks and splits in shingles with roofing cement.

Check the joints around flashing, which are common places for roof leaks to occur. Seal any gaps by cleaning out and replacing any failed roofing cement.

How to Replace Asphalt Shingles

Pull out damaged shingles, starting with the uppermost shingle in the damaged area. Be careful not to damage surrounding shingles that are still in good condition.

Remove old nails in and above the repair area using a flat pry bar. Patch damaged felt paper with roofing cement.

Install the replacement shingles, beginning with the lowest shingle in the repair area. Nail above the tab slots using ⅞ or 1" roofing nails.

Install all but the top shingle with nails, then apply roofing cement to the underside of the top shingle, above the seal line.

Slip the last shingle into place, under the overlapping shingle. Lift the shingles immediately above the repair area, and nail the top replacement shingle.

How to Replace Wood Shakes & Shingles

Split the damaged wood shingles or shakes with a wood chisel and mallet so they can be removed.

Remove the pieces. Slide a hacksaw blade under the overlapping shingles and cut the nail heads. Pry out the remaining pieces of the shakes or shingles.

Gently pry up, but don't remove, the shakes or shingles above the repair area. Cut new pieces for the lowest course, leaving a ⅜" gap between pieces. Nail replacements in place with ring-shank siding nails. Fill in all but the top course in the repair area.

Cut the shakes or shingles for the top course. Because the top course can't be nailed, use roofing cement to fasten the pieces in place. Apply a coat of roofing cement where the shakes or shingles will sit, then slip them beneath the overlapping pieces. Press down to seat them in the roofing cement.

How to Replace Valley Flashing

Asphalt roofing is most prone to leaking where two roof planes meet—in the valleys. Sometimes the metal valley flashing is corroded or poorly nailed, producing a leak; or if your shingles are woven over one another in a valley, there may not be any underlying metal flashing at all. As soon as the shingles start to deteriorate, a leak is bound to happen. It could also be that your current metal flashing's design is redirecting water back under the shingles during heavy rains and creating leaks. Whatever the case may be, you may need to replace the flashing or improve how it's installed. If the current shingles are in good condition, you can do this project without replacing your entire roof.

The process for replacing valley flashing involves four main stages. First, you'll carefully remove the shingles in the valley area so they can be reinstalled over the new flashing later. Second, self-adhesive underlayment membrane is rolled onto the roof deck to provide a barrier beneath the metal flashing. This step ensures that roofing felt isn't your roof's only line of defense if the metal leaks in the future. Third, you'll install new W-style aluminum valley flashing, which will never corrode. Finally, the original shingles are laced back into place over the flashing and cut back to leave the flashing area partially exposed.

Tools & Materials ▸

Pry bar
Push broom
Hook-bladed
    utility knife
Measuring tape
Straightedge
Hammer
Aviation snips

Metal seamer
Chalkline
Self-adhesive
    underlayment
    membrane
Aluminum W-style
    valley flashing
Roofing nails

Before

After

Damaged valley flashing can be patched for temporary leak stoppage, but you should replace it completely as soon as you can.

How to Replace Damaged Valley Flashing

Carefully lift shingles in the valley area with a prybar to break their self-seal strip, and pry the nails free. Shingles are generally held in place with eight nails—four from the shingle above and four above the tabs. Slide the shingles out and stack them in order so they'll be easier to replace later. Remove enough shingles to completely expose the valley flashing.

Pry off the old flashing and remove all nails. Sweep the valley area thoroughly, clearing off all of the debris on the roof deck. Inspect the roofing felt and decking surfaces. If the felt is torn or the decking is deteriorated from moisture, replace it now. (For more on replacing sheathing, see pages 42 to 43.)

Roll out the self-adhesive underlayment membrane from the ridge to the eave with the paper backing facing down. You can use granulated or bare membrane for this application. Cut the membrane off the roll so it overlaps the eave and ridge.

Starting at the ridge, peel off the backing paper to expose the adhesive and stick the membrane down on the roofing felt. You may find it helpful to tack the membrane at the ridge to anchor it. Slip the membrane under any overhanging shingles as you proceed, and keep the membrane as flat and smooth as possible. It must make full contact with the deck and not be gapped at the base of the valley, or it could tear and breach the seal.

(continued)

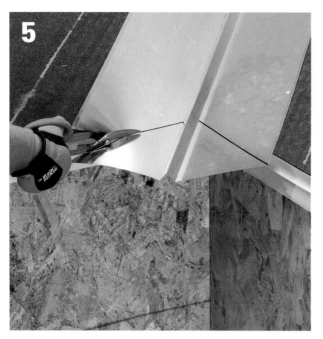

Set the first piece of valley flashing in place so it overlaps the eave. Use a straightedge to mark cutting lines on the flashing that match the angles of the roof decks. The goal here is to cut the flashing about 1" longer than the end of the roof so you can lap the flashing over the drip edge. It will prevent ice dams from forming here in the winter.

Use a metal seamer to bend the flashing overlaps down over the drip edge. Cut tabs in the notched area of the flashing and fold them over the raised ridge to close this gap.

Position the eave flashing carefully, and nail it in place with ⅞" roofing nails spaced every 12" up the flashing. Drive the nails within 1" of the edge of the flashing only.

When the first strip of flashing is completely nailed, set the second strip in place on the roof deck so it overlaps the first by 1 ft. With assistance, stretch a chalkline from the eave to the ridge to make sure both flashings form a straight line along the center ridge. Make any adjustments to the fit, and proceed to nail the second strip to the deck. Continue to install flashing up to the ridge, and cut off the overhang with aviation snips.

9

Slip the shingles back into position along one edge of the flashing from eave to ridge. Renail the shingles to the roof deck using the same nailing pattern as you would when installing them new.

10

Carefully lift the overlapping shingle edges, and nail the shingles to the valley flashing within 1" of its edge.

11

With an assistant's help, snap a chalkline from the ridge to the eave to mark a trim line on the overlapping shingles. Position this line so about 3" of the flashing next to the flashing ridge will be exposed at the roof ridge and about 6" will be exposed at the eave. The wider exposure of flashing at the bottom will help handle the greater volume of water here without overflowing the flashing.

12

Trim off the overhanging portion of each shingle along the chalkline to complete the first side of the flashing detail. Then repeat the process of refitting, nailing, marking, and trimming shingles on the other side of the flashing to finish the valley. Slip a scrap of wood behind the shingles to protect the flashing when trimming the shingles.

How to Replace Step Flashing

Carefully bend up the counter flashing or the siding covering the damaged flashing. Cut any roofing cement seals, and pull back the shingles. Use a flat pry bar to remove the damaged flashing. *Tip: When replacing flashing around masonry, such as a chimney, use copper or galvanized steel. Lime from mortar can corrode aluminum.*

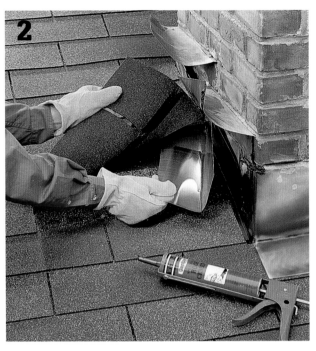

Cut the new flashing to fit, and apply roofing cement to all unexposed edges. Slip the flashing in place, making sure it's overlapped by the flashing above and overlaps the flashing and shingle below.

Drive one roofing nail through the flashing, at the bottom corner, and into the roof deck. Do not fasten the flashing to the vertical roof element, such as a chimney.

Reposition the shingles and counter flashing, and seal all joints with roofing cement.

Cleaning Roofs

If you live in a coastal region or on heavily wooded property, your asphalt roof may have discolored staining from mildew and algae growth. Provided these are just stains and not fungi or mold growth (signs of more serious moisture problems), you can remove the stains easily and have your roof looking new again. The process involves spraying the shingles with a solution of pool chlorine and liquid detergent, then rinsing it off with a garden hose. Chlorine will begin to kill the algae immediately and brighten the shingle color, but it will also continue to work for days after treating. Spray water onto shrubbery and the surrounding yard first, especially in downspout areas, to help dilute the chemical runoff.

Use extreme caution when walking on a wet roof. Wear soft-soled rubber sneakers for optimal traction, and protect yourself from the chlorine by wearing a rain suit and safety glasses. Choose a calm, overcast day for the job.

If your roof has a steep pitch, do not attempt this project; hire a professional roof-cleaning service to do the work instead.

Tools & Materials ▸

Ladder
Pump sprayer
Garden hose and
  spray nozzle

1 gal. of liquid
  pool chlorine
Liquid
  dish detergent

How to Clean an Asphalt Roof

Make the cleaner. Combine 1 gal. of swimming pool chlorine and 1 oz. of liquid dish detergent in a pump-style garden sprayer. Wear chemical gloves and safety glasses to protect yourself. *Tip: Spray your shrubs and yard with water to dilute any chemical overspray.*

Apply the cleaner. Working from the ridge to the eaves, apply the bleach solution with a garden sprayer in 6- to 10-ft. sections. Work quickly so the entire section can be rinsed clean while the treatment is still wet. Wear protective clothing, rubber-soled shoes, and safety glasses when working with the chemical spray.

Spray off the chemicals with a garden hose and clean water. Do not use a pressure washer for this step—it will blast granules off the shingles and shorten their life. Watch your step when working near the eaves. Follow up the first chemical spray and wash with a second application of a 50/50 mixture of chlorine and water. Rinse off the second treatment thoroughly with more fresh water before the chlorine dries.

SIDING

Basics of Siding

Few other exterior improvements give your home such a dramatic facelift as new siding. Siding allows you to change both the color and texture of outside walls. Siding products have come a long way in the past few decades, and your choices of colors and styles have never been greater. For a warmer, more rustic look, consider siding your home with cypress shakes or a cedar board and batten treatment, and then accent your foundation with veneered stone or brick. Wrapping a home with vinyl or wood is well within a seasoned do-it-yourselfer's capabilities. And when you're through, both you and the neighbors can appreciate your handiwork for many decades to come.

If you have a basic collection of hand and power tools, you've got most of the hardware needed for the job. Do a careful job of sheathing and wrapping your home to keep moisture outside where it belongs. Then get ready for a big project that will pay dividends in the end.

In this chapter:

- Choosing Siding
- Setting Up the Work Site
- Setting Up Scaffolding
- Pump-Jack Scaffolding
- Removing Siding
- Housewrap
- Vinyl Siding
- Lap Siding
- Wood Shakes & Shingles
- Plywood Siding
- Board & Batten Siding
- Log Cabin Siding
- Brickmold
- Finishing Walls with Masonry
- Brick
- Mortarless Brick Veneer
- Cast Veneer Stone
- Stucco
- Surface-bonding Cement
- Painting & Staining Siding
- Identifying Exterior Paint Problems
- Preparing to Paint
- Applying Paint & Primer
- Using Paint-Spraying Equipment
- Evaluating Siding & Trim
- Repairing Siding
- Repairing Stucco Walls

Choosing Siding

Replacing the siding on your home can be both more difficult and more expensive than adding a new roof. Siding is more challenging because of the many obstructions you have to work around—windows, doors, bumpouts, bay and bow windows, electric meters, phone and cable hookups, and more. In addition, your work site includes trees and bushes to work around—not the wide-open spaces of a rooftop. If you have a second story, you'll have to address the challenges of working with unwieldy pieces of siding at heights—and quickly discover that installing siding is best accomplished with a co-worker.

It can be more expensive for similar reasons. Window and door framing may have to be extended if you are adding new siding over old. Ancillary materials like house wrap, fasteners, caulk, and of course paint or stain, add up to a surprising degree. And you'll almost inevitably be using more material per square foot and pay more for it than you would for asphalt shingles.

Installation complexity: In general, the smaller the siding component, the easier time you'll have working around obstructions. For example, individual wood shingles are slower to apply than lap siding, but easy to finesse around windows and doors. Best of all, if you make a bad cut you've sacrificed only a single shingle and can back out of your mistake with little wasted material. At the other extreme, plywood siding covers a lot of ground quickly but requires more measuring and cutting skill to install. Make a mistake and you may be stuck with a useless chunk of plywood. Awkward materials can work to your advantage, however.

Cement siding is heavy, boosting its professional installation cost. That can be money in the pocket of a do-it-yourselfer willing to contribute some sweat equity.

Cost: Check the chart below for the relative costs of siding. Brick and stone are at the high end of the siding price scale, due both to material cost and the level of skill required for installation. Wood also tends to be pricey, especially the clear lumber used for lap siding or the wide cedar used in board and batten installations. Vinyl is cost effective but not everyone likes its appearance. Plywood—especially when trimmed out with vertical 1× strips for a board and batten look—is inexpensive, and if maintained, long lasting.

SIDING PRICE COMPARISON
Most Expensive
Brick
Stone
Stucco
Mid Range
Board & Batten
Wood lap
Wood shingles
Fiber cement
Least Expensive
Vinyl
Plywood

Maintenance Considerations

Maintenance tends to be more of a factor for siding than for roofing. The thought of having to paint your home's exterior every 7 to 10 years turns a lot of homeowners away from wood siding products. For others, painting is an opportunity to change the color of the house every decade or so. The amount of maintenance a siding material requires is not necessarily reflected in its cost, since products at both ends of the price scale—brick at the high end and vinyl at the low end—need little or no maintenance.

Older types of stucco were known to chip and crack and needed regular maintenance. Today's stucco products use epoxy and acrylics, greatly reducing such problems. If the final coat of stucco is tinted to the color you want, painting is unnecessary. Brick and stone sidings also require little maintenance, although the mortar joints may eventually need to be tuckpointed with new mortar. Likewise, vinyl siding is maintenance free, which is part of its appeal.

Wood siding products need the most attention. A coat of paint, stain, or sealer is usually needed every 7 to 10 years, a task that involves a lot of preparation work—scraping, sanding, and caulking. Fiber cement siding also needs to be repainted, but not as often; the standard time frame between paint jobs is 15 years. For both products, caulk is important not just to protect the substrate of your home, but also to keep wood from rotting and cement fiber siding from degrading. It is worth your while to annually check trim for any caulk failure.

Surprisingly enough, landscaping can have an affect on the life of your siding. Trees and bushes planted too close to the house harbor moisture that can discolor or damage your siding—even masonry. Cut back foliage at least a foot from siding. Check it once a year for regrowth.

Perhaps the greatest cause of siding damage is clogged or leaking gutters. Keep them cleared out and in good repair. Venture outside during a rainstorm to spot areas where gutters are sagging or downspouts loose. Both can cause a cascade of damaging water.

Combining the easy maintenance of vinyl with the good looks of shingles, vinyl shake siding is just one newer, low-upkeep siding option.

Siding Maintenance ▶

MATERIAL (SIDING)	MAINTENANCE	TYPE OF MAINTENANCE
Vinyl	Low	Occasional cleaning
Brick	Low	Filling mortar joints
Stone	Low	Filling mortar joints
Stucco	Low	Fix cracks/chips
Fiber cement	Medium	Painting
Wood shakes	Medium	Painting/staining
Wood shingles	Medium	Painting/staining
Wood lap	High	Painting

Vinyl & Metal Siding

Vinyl, aluminum, and steel siding were developed in the 1950s and 1960s and have been mass-produced for residential housing ever since. Chances are, most of the homes in your neighborhood are covered in one of these three types of siding. Consider the performance characteristics of each type when choosing new siding for your home.

Since siding is such a competitive market, the price of vinyl or metal siding is about the same. All three siding types are nailed in place along the top edge, with sections interlocking from one row to the next. Vinyl, aluminum, and steel can be installed by a do-it-yourselfer, but both aluminum and steel siding require the use of metal brakes and other sheet-metal tools to cut and form the siding correctly. Most homeowners find that installing metal siding is difficult to accomplish with acceptable results. Vinyl is much more DIY friendly. It can be cut easily with ordinary snips, circular saws, and utility knives.

A variety of special detailing components for wrapping windows, doors, and faucets or outlets are widely available at home centers.

Of the three types, vinyl siding dominates the residential siding market for several reasons. It is manufactured in hundreds of colors, so finding the "perfect" color for your home should be easy. The color is blended through the plastic, so scuffs or scratches will not leave noticeable blemishes. Vinyl is also molded in many surface textures and styles, including traditional or Dutch lap, scallops, shingles, fish scale, shakes, and various beaded designs. This diversity allows vinyl siding to complement historical homes and to provide design options for traditional and even eclectic homes. Vinyl is impervious to insects, rot, and fungal or algal growth, and it stands up well to moderate impact. It does not require routine cleaning or painting, and replacing sections of damaged vinyl is a fairly simple process.

Vinyl siding is very popular because it is inexpensive, low maintenance, easy to install, and widely available in many styles and colors (usually on the lighter side of the color spectrum). Vinyl is easy to cut and designed for foolproof installation (inset).

Metal siding is usually made of steel these days, but in the past, aluminum and tin have been common materials. Typically, it has a baked enamel finish. While a bit more challenging than vinyl siding, metal siding is also do-it-yourselfer friendly (inset).

The main drawback to vinyl is that it expands and contracts with changes in temperature more than other materials. If it's improperly nailed or installed without correct expansion gaps, the siding may buckle or sag. Vinyl also shows end seams from one piece to the next—and these can be unsightly, making a house look plasticized. Still, for durability and appearance's sake, vinyl is hard to beat. Most varieties of top-quality vinyl come with lifetime, transferable warranties (allowing the warranty to transfer from one homeowner to the next).

Aluminum and steel siding are manufactured in lap styles as well as several embossed surface patterns. Metal siding has a factory-applied, high-performance coating available in many colors. Steel siding is also galvanized for greater protection against corrosion. The rigidity of steel and aluminum siding ensures a strength that vinyl cannot match: steel and aluminum will not buckle, sag, or distort. Some manufacturers now offer seamless steel and aluminum siding, which is machine-extruded by the installer on location. Each panel can be custom fit to

wall lengths to avoid the staggered seam appearance of vinyl.

There are a few drawbacks to metal siding. Unlike vinyl, which is reasonably flexible and resilient to impacts, both steel and aluminum siding can be dented by the wayward baseball, rock, or falling tree branch. Since the surface of steel and aluminum is coated and the color does not permeate the metal, it is not resistant to scratches. Steel siding can rust if its cut edges are exposed to moisture. Sunlight can degrade the painted finish on aluminum siding over time, which leaves a chalky residue that occasionally needs washing. Both aluminum and steel siding may be painted, and you can repaint as needed without special preparation techniques.

Metal siding is a relatively green choice among the siding options. A percentage of the metal in most new steel or aluminum siding comes from recycled material and, depending on the finish coating, the siding may be recyclable when you replace it. Your supplier should be able to provide you with this information before you buy.

Cement Fiber Siding

Made of Portland cement, wood fiber, sand, and clay, cement fiber siding has boomed in popularity. Homeowners like its durability, modest cost, and standard 30-year warranty. It holds paint well, takes any impact you can throw at it, and has an exceptionally crisp look. It cannot rot, is termite proof, and if maintained can last even longer than masonry.

Lap siding paved the way, but manufacturers now produce cement fiber shingles, vertical siding for a board-and-batten look, and even trim and soffit components. Some manufacturers even offer baked-on finishes in a variety of colors.

Many homeowners find installing cement fiber siding challenging because the planks are heavy and prone to flop around if not carried exactly right. In addition, cutting them with a circular saw is extremely noisy and dusty. Two specialized tools help. Measuring gauges (see page 152) not only assist in keeping the exposure consistent, they act as a second pair of hands for holding the planks in place while fastening. Electric cutting shears neatly and quietly slice through siding with minimal dust.

Another advantage of cement fiber is its stiffness. Because it will not warp, it can be blind-nailed. That means the lower edge of each plank doesn't have to be nailed—the plank is attached by nailing only along its upper edge. Those nails are covered when the next course is installed.

If fiber cement has an Achilles heel, it is unpainted edges into which moisture can wick. It takes a long time, but eventually bare edges will swell. All cuts should be primed before installation and caulked. A spray can of primer is a handy way to coat cuts as soon as you complete them.

Cement fiber siding is best known as a lap siding alternative, but it is also available as shingles, board and batten, trim, and soffit panels. It is durable, long lasting, and relatively easy to install.

Wood Siding

Wood siding is another viable option for your home. Clapboard or shiplap siding is typically milled from cedar, cypress, redwood, or treated southern pine. All offer excellent resistance to decay and insects. Clapboard siding is usually manufactured with a tapered edge to create the proper overlap, and it's installed horizontally. Board-and-batten wood siding can be accomplished with ordinary lumber; it's arranged vertically with a narrower wood strip that overlaps the seams between the boards. Wood siding will generally cost more than vinyl or metal, and it's prone to the same problems as wood shingles and shakes. Prolonged wetness will eventually lead to decay, and the wood will fade to gray unless you're diligent about painting or staining every few years.

Wood lap siding is still a popular choice for homes, though vinyl, wood, aluminum, steel, and fiber cement reproductions are increasingly used. Wood is a satisfying material to work with, requiring only basic carpentry skills to install.

Shingles & Shakes

Wood shingles and shakes add a warm, natural look to your home. Wood has been used this way for centuries with excellent success. These days, shingles and shakes are made from various species of cedar and cypress; white oak; and pressure-treated yellow southern pine. The wood is graded in No. 1 and premium grades, based on clarity and cut. Premium is the highest grade. It costs slightly more than premium laminated asphalt shingles.

The principal difference between wood shakes and shingles is how they are cut and processed. Shakes are more rustic than shingles. They're sawn or hand split in varying thicknesses and lengths. These irregularities give shakes an organic appeal, but they increase the chances for water infiltration if the wall sheathing is not properly flashed and wrapped.

Shingles are manufactured to specific sizes and thicknesses, and they have a tapered edge. They are never hand split. Shingles can be ordered with sanded, smooth surfaces or left with the natural woodgrain texture.

A competent do-it-yourselfer can install shakes or shingles without special skills, and the material can be cut and installed with ordinary shop tools. Wood is abrasion resistant and easy to repair. However, the natural color will fade to gray over time unless it is painted. Shingles and shakes must be kept clean and dry to deter fungal or algal growth and insect pests. Woodpeckers and squirrels can wreak havoc with shingles and shakes when they're looking for a quick meal or a warm place to spend the winter. For all these reasons, you should expect to carry out more maintenance tasks on wood shingles and shakes than metal, vinyl, or other synthetic options.

Wood shingles and shakes can be used on the walls of your home as well as on roofs. Shakes have a more rustic appearance than shingles because they are irregular widths. Wood shingles do not install quickly but they are pleasurable to work with and make it easy to retrench if you make a mistake.

Panel Siding

For quick, cost-effective results, it's hard to beat panel siding (also called sheet siding). Panels are made of plywood, cement fiber, pressed hardboard, and OSB (Oriented Strand Board). The latter types are the least expensive but also the most prone to failure unless sealed completely with paint and caulk. However, all types of panels should be carefully sealed and kept maintained. Rough-sawn veneers mimic planks and are an ideal choice for achieving a board-and-batten look. Stain-grade veneers have no "football" patches. Panels come with or without grooves. Grooves are typically spaced every 4 or 8 inches.

Cement fiber panels are available in 4 × 8, 4 × 9, and 4 × 10 sheets. Wood types come in 4 × 8, 4 × 10, or 4 × 12 panels, each with a shiplap edge that overlaps its neighbor. The result is a neat and extremely weather-tight surface.

The chief drawback of panel siding is the difficulty of fitting them around windows and doors. A drywall T-square is a good aid for marking panels and a handy guide for sawing when clamped onto the panel. Applying substantial trim along the bottom of each wall greatly simplifies installation because you can set a panel on the trim and tip it into place for fastening. In addition, the trim protects the edge of the panel from ground splash.

Whichever type of panel you choose, be sure to add flashing above windows and doors, and when one panel is placed above another. Unless your panels are pre-primed, go to the trouble painting edges in advance with exterior-grade primer. Once you get rolling, spray primer is a quick way to safeguard saw cuts. In addition, remember that a tight fit is not always best. Follow manufacturer's instructions for leaving an expansion gap between panels.

Simple but elegant, panel siding is the quickest type of siding to install. These cement fiber panels were prepainted for a luminous, wood-like effect.

Board & Batten Siding

For a traditional look, a board and batten appearance can be recreated with 4 × 8 sheets of exterior-rated plywood that have a textured surface and a pattern of grooves to simulate the look of individual boards (called T1-11). Grooves are spaced either 4 or 8 inches apart. The edges of the sheets have overlapping tongues that hide the seams and help keep water out. Installation is easy for a competent homeowner with basic woodworking skills, and the material costs are about the same as vinyl siding.

Board and batten paneling must be carefully installed and properly flashed so the back surfaces and edges remain dry. It also needs a protective topcoat of paint or stain to prevent deterioration from moisture or sunlight. As the panels age, it's common for the laminated layers to begin to separate. But, when the sheets eventually wear out, they're easy to replace.

You can also achieve the board and batten look by attaching narrow strips of wood, such as 1 × 2, vertically over seams between sheets of textured paneling, and also at regular intervals in the field area of the panels.

Board and batten siding was traditionally made up of wide, solid wood boards attached in a vertical configuration with narrow wood strips attached to conceal the gaps between boards. Today, this look is frequently recreated with plywood panels.

T1-11 paneling approximates the look of traditional board and batten siding, it is easy to install, and it is inexpensive. This paneling is still used for home siding, but you'll see it more often on sheds and outbuildings these days.

Log Cabin Siding

Log cabin siding is yet another wood siding alternative. Typically, white pine or cedar logs are stripped of bark, and then sawed into thin lengths of siding with the curvature of the log's surface left intact. The curvature can also be created at the mill by machines. The siding is installed with nails over building felt or housewrap, just like other siding options. The results are surprisingly convincing. It takes close inspection to realize that walls aren't made of whole logs. Any competent do-it-yourselfer can successfully install log cabin siding with common hand and power tools, and repairs are easy to make. Log cabin siding needs a top coat like other wood siding options to extend its serviceable life. Of course, log siding is also prone to degradation from insects, other animal pests, and rot. It's priced comparably to lap siding or shingles.

You don't have to be a lumberjack to live in a log cabin home. Here, the rustic look of hewn logs and interlocking log-tail corners belies the fact that it is actually wood siding installed with a hammer and nails. It's a perfect do-it-yourself invitation to alpine living.

Masonry, Stone & Stucco

Various cement-based products give you many more options for durable, long-lasting siding. You're probably already familiar with stucco, which is a mixture of Portland and masonry cements, sand, and water. Stucco is installed in successive coats over a base of felt and metal lath. It can be applied to either wood-sheathed or masonry walls. A variety of colorants can be added to the finish coat. Alternatively, stucco can be painted with specially formulated elastomeric paints. Texturing is also a common practice, offering a blended, more natural appearance.

When properly installed, stucco will last for decades with only an occasional restorative coat. It can crumble or crack as a house settles, but stucco generally proves to be a tough, low-maintenance exterior finish. It may be used to coat entire exterior walls or for limited areas in conjunction with other siding materials. Removing stucco or creating new openings for windows and so forth is possible but labor-intensive.

Bonding cement can also be used to cover exterior walls. Essentially, it is a blend of Portland cement, sand, and fiberglass combined with an acrylic fortifier to create a cementitious plaster. Bonding cement is often applied in a smooth layer to hide cinder block or brick foundation surfaces. Because it isn't enhanced with color or texture, it's used sparingly to detail other forms of siding.

Mortared-in-place natural stone or full bricks are other attractive siding treatments to consider. Both are heavy and may require reinforced walls. The costs of quarrying and transporting natural stone make it expensive, but it's hard to argue against its beauty. Natural stone can be used sparingly to enhance foundation walls or pillars. It is common to combine rock with other forms of siding to help limit costs.

Conventional bricks provide a rugged or refined elegance, depending on the style you choose. They are durable, immune to insect damage, and fireproof. Brick repairs are considerably more difficult, and creating new openings for windows or doors can be labor-intensive. When installed correctly, stone and brick will last the lifetime of your home.

Stucco has a long tradition as a home siding material, especially in areas of the country where wood products are scarce. It is a very durable siding material, but it does require some maintenance.

Brick veneer lends a stately appearance to ordinary-looking homes. It does help keep the elements out, but it does not have any structural value.

Stone veneer is a home design element that is on the rise in popularity. Today, more likely than not, the veneer you see is a lightweight, manufactured product rather than genuine stone.

Mortarless stone veneer siding is affixed to the facade of a house using only mechanical fasteners and supports. It provides the beauty of real masonry without the trouble of mixing and tooling mortar.

If you want the look of stone or brick on your house without the rigors of cutting and moving heavy materials, consider using manufactured stone and brick veneer. These products are made of cement that's cast into natural stone forms or brick shapes.

A multitude of simulated stone styles are available, including round, flat, textured, and smooth. Color options vary widely as well. Veneered stone is fixed in place with mortar over a base of felt and wire mesh. It weighs a fraction of real stone, is more affordable, and the end result is very convincing.

A mortarless system of brick veneer is often a better choice for do-it-yourselfers. Bricks simply stack in interlocking courses on the surface of a sheathed and wrapped wall. There is no mortar because every fourth course is fastened with screws to furring strips.

Stucco, veneered stone, and mortarless brick can be installed with moderate masonry skills and some specialized tools. Fiber cement siding is also user-friendly, but you'll need electric shears and carbide blades on other tools. Hire a professional mason to install mortared, natural stone, or full brick.

Setting Up the Work Site

Allow at least half a day to pull your tools together and prep your site. To plan the site, take some time to think through your workflow. For example, you'll want a temporary staging area where you can store a day's worth of siding close to your cutting area but not underfoot. Position the cutting area convenient to the side of the house you are working on. Plan on moving it as you progress. Haul in a garbage can or have a small dumpster dropped off so you can keep up with the debris—too much junk lying about is a tripping hazard. Reserve an area for cutoffs that may prove handy later. If you have a wheelbarrow, consider making it your permanent repository for nails, caulk, and other incidentals. You'll find it handy for wheeling on and off the job site each day.

Obstructing bushes are one of the greatest irritants when you are installing starter courses close to ground level. Prune them in advance—their foliage should be no closer than a foot from your house anyway. Tie bushes back with rope, staking them so they're pulled away from the house.

Use at least two power cords—more if possible. You'll save yourself a lot of irritation if you can plug in all your tools at once—it is no fun getting all ready to make a cut with your miter saw only to find that it is unplugged. Having tools near at hand makes them less likely to get lost. Improvise a tool table with a sheet of plywood on a couple of sawhorses. It will spare you frequent bending over.

All of this may seem like a delay of game, but it's not. Get things set up for action in advance and you'll do better work and enjoy it more.

Setting Up Ladder Jacks ▸

Attach a ladder jack to your ladder by slipping the rung mounts over the ladder rungs. Level the platform body arm, and lock it into place.

Set the plank in place on the platform arm. Adjust the arm's end stop to hold the plank in place.

Ordering Siding Materials

Siding is sold by the linear foot, square foot, board foot, or square, depending on the type of material. To determine the amount of siding materials you'll need, calculate the square footage of each wall, then add them together to get your total surface area.

To figure the square footage for a wall, multiply the length of the wall by the height. Calculate the area on the gable end of a wall by using the formula for triangles, which is one-half the length of the base multiplied by the height of the triangle. You won't need to subtract for areas covered by windows and doors, and you won't need to add an extra 10 percent for waste. The window and door areas are roughly equal to the amount of waste you'll need.

Use your height and length measurements from the walls to determine how many feet of starter strip, channels, corner posts, and trim you'll need. Keep in mind you'll probably also need to apply trim or channels around all of your doors and windows.

Depending on the siding material you're ordering, you may need extra material to allow for overlap. For example, if you want to install 10" lap siding with an 8" exposure, you'll need to account for a 2" overlap for each board. Likewise, the exposure rate for wood shakes or shingles will determine how many squares you'll need. Depending on the exposure rate, one square of shingles could cover 80 sq. ft., 100 sq. ft., or even 120 sq. ft. Most manufacturers have charts that show how much material is needed to cover a specified number of square feet at various rates of exposure. If you have trouble estimating how much material you need to purchase, ask your supplier for help.

Estimate the amount of siding you'll need by calculating the square footage of each wall, then adding the numbers together. To determine the square footage of a wall, multiply half the length of its base by its height. Adding an extra 10 percent to the answer to account for waste isn't necessary. Don't forget about corner trim, J-channel and other trim pieces.

Preparing the Job Site

Cover external air conditioning units, appliances, and other structures near the house. Make sure the power is turned off before covering them, and don't use them while they're covered.

Remove exterior shutters and decorative trim to protect them from damage and make it easier to paint.

Create a tool platform with sawhorses and a piece of plywood. Tools spread on the ground are a safety hazard and moisture can damage them. It's safer and more efficient to organize your tools so they're dry and easy to find.

Setting Up a Saw Area

Neat cuts—and your personal safety—depend on a sawing setup that supports your work before and after the cut. The ideal arrangement is at a comfortable working height (adjustable saw horses help), with plenty of elbow room for hefting siding in place.

Beg, borrow, or steal extra sawhorses. For any sort of plank siding, a 12 ft. 2 × 12 supported by sawhorses makes a stable base for cutting with a circular saw. That way, you can support your work with scraps of 1 × 2 (see below).

A miter saw needs a small table. Often it makes sense to permanently attach it to a 2 × 4-ft. piece of ¾" plywood so you can set it directly on sawhorses. Other sawhorses, bolstered by clamped-on scraps of wood, support long planks while you are cutting.

Four 2 × 4s set on sawhorses are an easy way to support a panel while cutting. Arrange your work so that when the cut is finished, both pieces are fully supported. A drywall T-square is handy for marking and, when clamped in place, serves as a cutting guide.

Support long panels with 2 × 4s underneath to eliminate any binding while sawing. Clamp a straightedge or drywall T-square as a guide while cutting longer panels or lumber.

Long planks need a lanky base when sawing; a 2 × 12 is heavy enough to do the job without sliding around. Have at least four 1 × 2 scraps handy to use as spacers that support both sides of your work.

Setting Up Scaffolding

A set of ladders and roof jacks may be all the support you need for roofing projects, but replacing siding higher than the first floor of your home will require some type of platform-style scaffolding. It isn't safe to slide a plank between two ladders and call it good; the setup lacks stability and protection from swaying. A broad platform is more helpful for staging materials and tools and moving around freely.

Steel tubular scaffolding is available from any rental center for daily, weekly, or monthly rates. It should come with a complete set of instructions for how to assemble the pieces correctly. Begin by setting up the scaffolding on flat ground, free from mud and construction debris. If you have to adjust the first level for uneven terrain, follow the instructions carefully. Do not use stacks of boards or cinder blocks to create a level surface.

Each stage of scaffolding consists of two end frames, several crossbraces, and planking that makes up the platform. Set up the first level of scaffolding on steel base plates, if your equipment has them, or on wide wood base supports. Use string lines and bubble levels to make sure the end frames are level and plumb. If the scaffolding has screw jack adjusters, use them to level the structure.

Make sure all the components are fully seated in their joints, then add the planking to create the first platform. The platform shouldn't sway. Platform planks should line up evenly to prevent tripping hazards. Keep gaps between the planks to 1 inch or less. Once the first level of staging is secure, repeat the process to build the second stage of scaffold. Make sure any connective locking pins are securely engaged from one stage to the next. If you need to build scaffolding more than 16 feet high, use wall brackets that clamp to the scaffold and bolt them into solid wall framing for added stability. No matter how high your scaffold becomes, it must remain as stable as the first stage. Add any guard rails, end rails, or toe boards that may be required.

Renting scaffolding is an excellent investment if you are painting or installing siding. You easily recoup the cost in time savings and safety.

Securing Scaffolding ▶

Sections of scaffolding are connected with locking cotter pins that fit through holes aligned in the tubular legs. The cotter pins feature clips that snap around both ends to keep them from slipping out of the tubular legs. Even with these locking pins it's always a good idea to test the connections daily before using your scaffolding.

Tools & Materials ▶

Scaffolding components
String line
Levels
Shovel
(for leveling the installation area)

How to Set Up Scaffolding

Clear the setup area of debris, and then assemble the first two end frames and crossbraces to create the bottom stage of scaffold framing.

Use string lines and levels to check end frames for level and plumb. Adjust the structure using screw jacks mounted to the frame legs, or by a leveling technique recommended by the manufacturer. The end-frame legs should rest firmly on wood or steel base plates. Never use stacks of boards or cinder blocks for leveling purposes. Lay the planks in position to form the bottom platform.

Assemble the parts for the second stage on top of the first stage. Make sure all joints and connective locking pins are fully engaged from one stage to the next (see tip, previous page).

For extreme heights that require more than two stages, you need to secure the scaffolding to the house with chains or wall tie-in brackets that clamp to the scaffolding. Make sure to fasten to wall framing—not to sheathing only.

Pump-Jack Scaffolding

Now and then you may see professional siding crews using an alternative form of scaffolding called pump-jack scaffolding. Pump jacks consist of long wood or aluminum support posts that extend from the ground to the roof line. A foot-operated jack mounts on each pole and supports wooden or aluminum walk planks. Some designs also include a secondary platform that can be used as a workbench or serve as a guardrail. The advantage to using pump jacks over staged scaffolding is adjustability. By pumping the foot lever on each mounting post, workers can raise the platform in 6-inch intervals. Pump jacks are particularly well suited for siding and painting projects, which require moving up the walls incrementally. Staged scaffolding offers a larger platform, but it's set at a fixed height.

Pump jacks are less common than staged scaffolding at rental centers. Siding contractors lend them out during the off-season as well. With some research, you can find them for $150 a pair. While pump jacks are stable and versatile, they don't offer the same level of fall protection as a full framework of scaffolding, and you won't have a great deal of room to place extra project materials and tools. If you have access to fall-arresting gear, wear it.

Tools & Materials ▸

Hammer
16d nails
Pump jack
   components

Wood support posts
   (if applicable)
Fall-arresting gear
   (highly recommended)

Pump jacks are a convenient alternative to scaffolding. They offer greater foot space and hence are safer than ladders. The steel jacks that support a cross-plank are raised and lowered on a pole or post by foot operation.

How to Use Pump-Jack Scaffolding

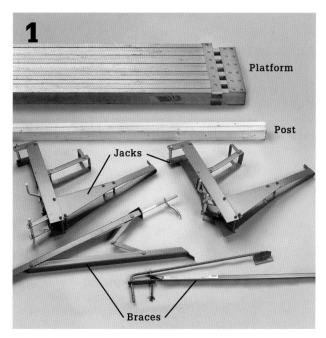

Assemble the components. You'll need a pair of foot-operated pumps that raise a work platform along wood or aluminum poles. The components assemble quickly and provide more flexibility than standard scaffolding for working at varying heights. They're ideal for siding installation.

Build the support posts. Depending on the pump jacks you use, you may need to first build the posts from common 2 × 4s and 16d nails. Wood posts can be made up to 30 ft. tall, provided the joints are staggered and mending plates are nailed over every joint.

Secure the posts to the framing members of the exterior wall using triangular pump jack post brackets.

Install the platform and adjust the height as needed. Two operators can raise or lower the work platform quickly, but only one person can operate one pump at a time. Plan your material quantities and tool requirements carefully before raising the platform. Space is more restricted on the platform, so it must be lowered down to replenish tools and materials.

Removing Siding

Although it's sometimes possible to install new siding over old if the old siding is solid and firmly attached to the house, in most cases it's better to remove the siding, especially if it's damaged. Taking off the old siding allows you to start with a flat, smooth surface. And because the overall thickness of the siding will remain unchanged, you won't have to add extensions to your window and door jambs.

There's no "right" way to remove siding. Each type of siding material is installed differently, and consequently, they have different removal techniques. A couple of universal rules do apply, however. Start by removing trim that's placed over the siding, and work from the top down. Siding is usually installed from the bottom up, and working in the opposite direction makes removal much easier. Determine the best removal method for your project based on your type of siding.

Strip one side of the house at a time, then re-side that wall before ripping the siding off another section. This minimizes the amount of time your bare walls are exposed to the elements. Take care not to damage the sheathing. If you can't avoid tearing the housewrap, it can easily be replaced, but the sheathing is another story.

While the goal is to remove the siding as quickly as possible, it's also important to work safely. Take care when working around windows so the siding doesn't damage the glass. Invest the necessary time to protect the flowers and shrubs before starting the tear off (page 29).

Renting a dumpster will expedite the cleanup process. It's much easier to dispose of the siding as soon as it's removed rather than stacking it up in an unsightly pile in your yard, then throwing it away later. When you're finished with your cleanup, use a release magnet to collect the nails on the ground.

Tools & Materials ▶

Cat's paw	Masonry-cutting
Flat pry bar	blade
Zip tool	Masonry bit
Drill	Aviation snips
Circular saw	Roofing shovel
Masonry chisel	Release magnet
Hammer	

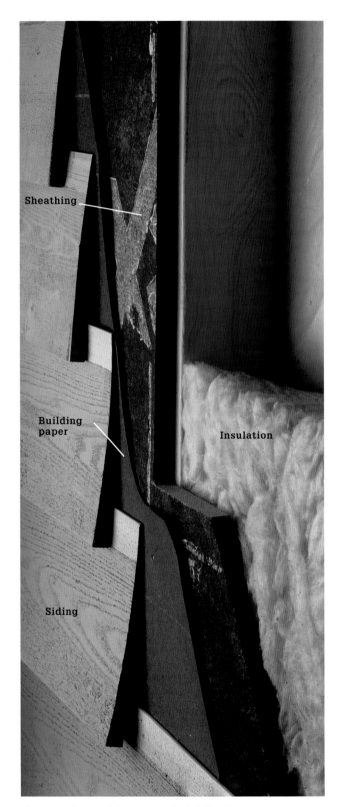

The exterior wall is composed of siding, housewrap or felt paper, and sheathing. Try to remove the siding without disturbing or damaging the sheathing.

Tips for Removing Siding

Brick molding comes preattached to most wood-frame window and door units. To remove the molding, pry along the outside of the frame to avoid marring the exposed parts of the jambs and molding.

Lap siding is nailed at the top, then covered by the next course. Pry off the trim at the top of the wall to expose the nails in the top row. Remove the nails using a cat's paw, and work your way down the wall.

Shakes and shingles are best removed with a roofing shovel. Use the shovel to pry the siding away from the wall. Once the siding is removed, use the shovel or a hammer to pull out the remaining nails.

Board and batten siding is removed by prying off the battens from over the boards. Use a pry bar or cat's paw to remove the nails from the boards.

Siding shown cutaway for clarity

Vinyl siding has a locking channel that fits over the nailing strip of the underlying piece. To remove, use a zip tool to separate the panels, and use a flat pry bar or hammer to remove the nails.

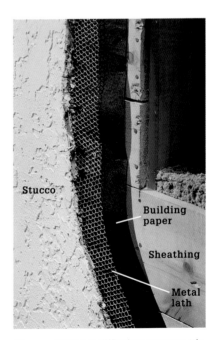

Stucco

Building paper

Sheathing

Metal lath

Stucco siding is difficult to remove. It's usually much easier to apply the new siding over the stucco than to remove it. If you're determined to take it off, use a cold chisel and hammer to break it into pieces, and use aviation snips to cut the lath.

Housewrap

Housewrap is a specially engineered fabric that blocks air and water infiltration from the outside but allows moisture vapor to pass through from the inside. It's best to apply the housewrap before installing windows and doors, but since that's not always possible with a remodeling or siding replacement job, you can cut the housewrap to fit around them. Most siding materials need to be nailed to studs, and the marks on the housewrap identify their locations. Staples are permissible for fastening housewrap, but cap nails are recommended and have better holding power.

Felt paper is not the same as housewrap. It's not necessarily designed to work as an air barrier, and it may absorb water. Do not substitute felt paper when housewrap is supposed to be used.

Wall sheathing stiffens building wall framing and creates a uniform backing for siding and trim. A layer of building paper or housewrap seals the sheathing from moisture infiltration.

How to Install Housewrap

To install housewrap, begin at the bottom courses if the product you're using is not wide enough to cover a wall in one piece. *Note: Housewrap is a one-way permeable fabric that helps keep moisture from entering the structure from the exterior. Installing it only makes sense when planning to finish the interior walls of the house or structure.*

Attach the housewrap with housewrap nails. Drive the nails every 16" into each wall stud and horizontal framing member.

When starting a new roll, overlap vertical seams by 6 to 12", aligning the stud marks. Begin the second course by overlapping the bottom course by 6". Once again, make sure stud marks are lined up.

Finish installing the housewrap. All seams should overlap by at least 6 to 12", with horizontal seams overlapping from above.

Cut out windows and doors. Make a long X cut in the housewrap, connecting corners diagonally at window and door openings. Use a utility knife to make the cut. Staple down the extra housewrap in the window rough opening so it wraps around the jack studs, header, and rough sill.

Tape the seams. To seal the housewrap, apply housewrap tape along all horizontal and vertical seams. *Note: Housewrap is not rated for long-term exposure to the sun, so do not wait more than a few weeks after installing it before siding the structure.*

Vinyl Siding

Vinyl has become one of the most popular sidings due to its low cost, uniform appearance, and maintenance-free durability. Installation is fairly simple, with each row locking onto the lip of the lower course, then nailed along the top.

There are a couple of key factors that will make or break your siding project. First, the sheathing must be straight and solid before the siding is applied. The siding will only look as straight and smooth as the wall it's on. Second, determine how the siding should overlap to hide the seams from the main traffic patterns. This usually means starting in the back and working toward the front of the house.

Do not nail the siding tight to the house. The panels need to slide back and forth as they expand and contract with changes in the temperature. If the siding can't move, it will bow and need to be reinstalled. Keep a ¹⁄₃₂" gap between the head of the nail and the siding.

Vinyl siding is available in a wide variety of colors and styles, and with a lot of accessories such as trim, fluted lineals, vertical columns, crown molding, and band boards. The most common vinyl siding is horizontal lap siding, which is shown starting on page 140. This project shows a foam underlayment, which reduces outside noise, protects the siding from dents, and adds an insulation value. Vertical vinyl siding is also available. It's specifically made for vertical applications.

Tools & Materials ▸

Hammer	Zip-lock tool
Circular saw	Snap lock punch
Miter saw	Caulk gun
Clamps	Vinyl siding
Tape measure	J-channel
String	Corner posts
Straightedge	Undersill
Utility knife	Starter strip
Aviation snips	Nails
Level	Cutting table
Chalkline	Safety glasses
Framing square	Silicone caulk
Nail slot punch	

Vinyl siding can look very similar to wood lap siding, but it doesn't require regular upkeep. Vinyl can be installed on any type and style of house.

Tips for Cutting Vinyl Siding

Use a fine-tooth blade installed backward in the saw to cut vinyl siding. Use a radial-arm saw, power miter saw, or a circular saw, and move the blade slowly through the siding. Always wear safety glasses when cutting siding.

Support the siding on a cutting table when cutting. Vinyl siding is too flimsy to be placed across sawhorses without support. You can build a cutting table by fastening a long piece of scrap plywood between two sawhorses.

Tips for Using Specialty Tools

A snap lock punch is used to make raised tabs, or dimples, in a cut edge of siding where the nailing hem has been removed. This eliminates the need to facenail the panel.

A nail slot punch is used to make horizontal nail slots in the face of panels. It can also be used to add or elongate the opening of an existing nail slot to match irregular stud spacing.

How to Install Vinyl Siding

Install housewrap following instructions on pages 136 to 137. Identify the lowest corner of the house that has sheathing, and partially drive a nail 1½" above the bottom edge of the sheathing. Run a level string to the opposite corner of the wall and partially drive a nail. Do this around the entire house. Snap chalklines between the nails.

Place the top edge of the starter strip along the chalkline and nail every 10". Nail in the center of the slots and don't nail tight to the house. Keep a ¼" gap between strips, and leave space at the corners for a ½" gap between starter strips and corner posts.

Option: Install foam vinyl siding underlayment on the house using cap nails. Align the bottom of the underlayment with the starting strip. To cut panels to size, score them with a utility knife, then break them over your cutting table. Some panels need to be taped at the seams. Follow manufacturer's recommendations.

Install a corner post, keeping a ¼" gap between the top of the post and the soffit. Extend the bottom of the post ¼" below the bottom of the starter strip. Drive a nail at the top end of the uppermost slot on each side of the post (the post hangs from these nails). Make sure the post is plumb on both sides using a level. Secure the post by driving nails every 8 to 12" in the center of the slots. Do not nail the post tight. Install the other posts the same way.

If more than one corner post is needed to span the length of a corner, the upper post overlaps the lower post. For an outside corner post, cut off 1" from the nailing flanges on the bottom edge of the top post. For an inside corner post, cut off 1" from the nailing flange on the upper edge of the bottom post. Overlap the posts by ¾", leaving ¼" for expansion.

Measure, cut, and install undersill along the horizontal eaves on the house. If more than one undersill is needed, cut the nailing hem 1¼" from the end of one undersill. Overlap the undersills by 1".

Measure and cut two J-channels that are the length of a window plus the width of the J-channel. Place one of the J-channels against the side of the window, aligning the bottom edge with the bottom edge of the window. Nail the channel in place. Nail the second J-channel against the opposite side of the window the same way.

At the top of the window, measure between the outside edges of the side J-channels and cut a piece of J-channel to fit. Cut a ¾" tab at each end. Bend the tabs down to form a drip edge. Miter cut the face at each end at 45°. Center the J-channel over the window and nail it in place. The top J-channel overlaps the side pieces, and the drip edges fit inside the side pieces. Do this for each window and door.

(continued)

Measure, cut, and install undersill beneath each window. The undersill should be flush with the outside lip of the side channels.

Measure, cut, and install J-channel along the gable ends. Nail the channels every 8 to 12". To overlap J-channels, cut 1" from the nailing hem. Overlap the channels ¾", leaving ¼" for expansion. At the gable peak, cut one channel at an angle to butt against the peak. Miter the channel on the opposite side to overlap the first channel.

To install J-channel over a roof line, snap a chalkline along the roof flashing ½" above the roof. Align the bottom edge of the J-channel along the chalkline, and nail the channel in place. Make sure the channel does not make direct contact with the shingles.

Snap the locking leg on the bottom of the first panel onto the starter strip, making sure it's securely locked in place. Keep a ¼" gap between the end of the panel and the inside of the corner post. Nail the panel a minimum of every 16" on center. Don't drive the nails tight. *Note: This installation shows a vinyl siding underlayment in place.*

12

Overlap panels by 1". Cut panels so the factory cut edge is the one that's visible. Keep nails at least 6" from the end of panels to allow for smooth overlap. Do not overlap panels directly under a window.

13

Place the second row over the first, snapping the locking leg into the lock of the underlying panels. Leave ¼" gap at corners and J-channels. Install subsequent rows, staggering seams at least 24" unless separated by more than three rows. Check every several rows for level. Make adjustments in slight increments, if necessary.

14

For hose spigots, pipes, and other protrusions, create a seam at the obstacle. Begin with a new panel to avoid extra seams. Cut an opening ¼" larger than the obstacle, planning for a 1" overlap of siding. Match the shape and contour as closely as possible. Fit the panels together around the obstruction and nail in place.

15

Place mounting blocks around outlets, lights, and doorbells. Assemble the base around the fixture, making sure it's level, and nail in place. Install siding panels, cutting them to fit around the mounting block with a ¼" gap on each side. Fasten the cover by snapping it over the block.

(continued)

16

Where panels must be notched to fit below a window, position the panel below the window and mark the edges of the window, allowing for a ¼" gap. Place a scrap piece of siding alongside the window and mark the depth of the notch, keeping a ¼" gap. Transfer the measurement to the panel, mark the notch, and cut it out. Create tabs on the outside face every 6" using a snap lock punch. Install the panel, locking the tabs into the undersill.

17

Install cut panels between windows and between windows and corners as you would regular panels. Avoid overlapping panels and creating seams in small spaces. The panels need to align with panels on the opposite side of the window.

18

To fit siding over a window, hold the panel in place over the window and mark it. Use a scrap piece of siding to mark the depth of the cut. Transfer the measurement to the full panel and cut the opening. Fit the cut edge into the J-channel above the window, lock the panel in place, and nail it.

19

For dormers, measure up from the bottom of the J-channel the height of a panel and make a mark. Measure across to the opposite J-channel. Use this measurement to mark and cut the panel to size. Cut and install panels for the rest of the dormer the same way.

Measure the distance between the lock on the last fully installed panel and the top of the undersill under the horizontal eaves. Subtract ¼", then mark and rip a panel to fit. Use a snap lock punch to punch tabs on the outside face every 6". Install the panel, locking the tabs into the undersill.

Place a scrap panel in the J-channel along the gable end of the house. Place another scrap over the last row of panels before the gable starts, slide it under the first scrap, and mark the angle where they intersect. Transfer this angle to full panels. Make a similar template for the other side. Cut the panels and set the cut edge into the J-channel, leaving a ¼" gap.

Cut the last piece of siding to fit the gable peak. Drive a single aluminum or stainless steel finish nail through the top of the panel to hold it in place. This is the only place where you will facenail the siding.

Apply caulk between all windows and J-channel, and between doors and J-channel.

Lap Siding

There are several types of horizontal board siding applications, such as clapboard, tongue and groove, bevel, and shiplap, but the most popular is lap siding, which is the project we're showing here. The installation is fairly straightforward, with each course overlapping the underlying row and covering up the nailing. We're using measuring gauges that hold both ends of a course in place, making installation simpler and easier (see page 152).

Before installing wood siding, make sure it's acclimated to your environment so it can expand or shrink prior to being nailed in place. For our project, we're using fiber-cement siding, which offers the look and texture of wood but doesn't rot or crack. The application is basically the same, although fiber cement can be more difficult to nail (you may want to predrill holes) and requires a carbide-tipped blade for cutting. Wear a respirator when cutting fiber cement since it contains silica, which can cause lung disease.

Store siding in a flat position, and keep it off the ground and covered until ready for use. When carrying the siding, hold it edge-up like you would a book to make sure it doesn't bend or crack. Having a helper makes this job much easier. The siding panels need to be nailed to studs, so it's critical for the marks on the housewrap to be aligned with the wall studs. Cut the panels face down to avoid marring or damaging the faces.

If the siding is not yet primed, apply a coat of primer before installing. Also apply primer to cut edges during the installation process. Wood and most fiber-cement siding needs to be painted after it's installed. Although you'll have a paint job on your hands when the project is finished, the upside is you'll have the opportunity to change the color of your siding whenever you want by applying a new coat of paint. Other sidings are available in colors that don't require painting, and the seams are caulked with a matching colored caulk.

Lap siding can be wood, fiber-cement, or one of many manufactured materials. It is easy to install and has a traditional appearance.

Tools & Materials ▸

Tape measure	T-bevel
Circular saw	¼ × 1½" lath
Caulk gun	Siding and trim
Chalkline	10d corrosion-resistant
Paintbrush	casing nails
Combination saw	2" corrosion-resistant
blade (for wood)	siding nails
Carbide-tipped	Flexible caulk
saw blade	Primer
(for fiber cement)	Power shears (for fiber
4-ft. level	cement)
Measuring gauge	Story pole

Cutting Techniques for Lap Siding

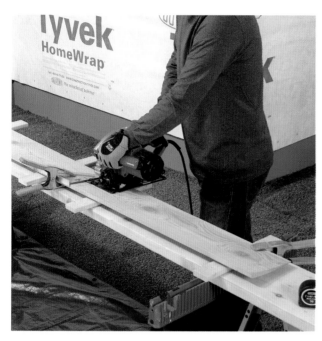

Cross cuts with a circular saw are made simpler with the use of a small square. Always start the saw before pushing it into the material. With the square clamped to the siding and both hands on the saw, begin the cut at your mark and follow through with one smooth, continuous motion, using the square as a guide.

For ease and speed use a miter saw. While the saw can't do plunge and rip cuts, it can't be beat for quick, accurate cross cuts. Often you'll have to cut several planks exactly the same length—clamping them together before cutting saves time and boosts accuracy.

A rip cut can be challenging but is made easier if your work is supported. For a straight cut, use a saw equipped with a guide. You may also clamp a straightedge to your plank; if your cut ends up with some wobbles, a few strokes with a block plane will smooth it.

A scarf joint helps diminish any gaps because of the angled cut used. While the angles can be a bit tricky to manage (inset), a scarf joint can be cut with either a compound miter saw or a circular saw.

Cutting for Obstructions

Use a jigsaw to cut holes. Mark the opening and bore access holes in two opposite corners with a ½" bit. Slip the jigsaw blade into each hole to cut out the sides.

For a neat installation of siding around a hose bib, go to the trouble of turning off your household supply and removing the hose bib. If adding new siding over old, you may have to extend the supply line with a short length of pipe called a nipple.

Use a plunge cut to make a notch in siding above or below a window. It takes a steady eye to get this right, so if you aren't familiar with plunge cuts, practice a few first. Watch the blade, not the guide notch of the base. Slowly push the saw forward as you lower the blade into the siding.

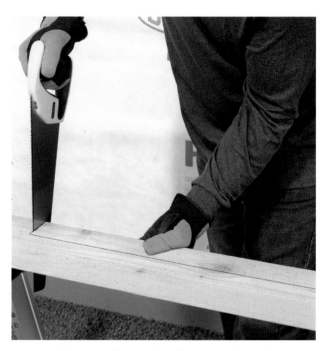

To complete a notch, use a handsaw to cut through whatever the circular saw blade can't reach. Holding the saw backwards makes it easy to hold it vertical to the plank while you complete the cut.

How to Install Lap Siding

Cover the exterior walls with housewrap so the stud marks fall on the studs (see pages 136 to 137). Starting at the lowest corner of the house, snap a level line at the bottom of the wall where the siding will begin. The siding should cover the sill plate and just lap over block or concrete, but stay above grade.

Install a corner trim board flush with the outside wall and flush with the chalkline at the bottom. Keep nails 1" from each end and ½" from the edges. Drive two nails every 16". Overlap a second trim board on the adjacent side, aligning the edge with the face of the first board, and nail in place.

When two or more corner trim pieces are needed to complete a wall, cut a 45° bevel on the end of each board. Apply primer to the cut ends. Install the first board so the bevel faces away from the house. Place the top piece over the first board, aligning the bevels. Stagger seams between adjacent sides.

Place a corner trim board in an inside corner. Drive nails every 16".

(continued)

Corner Trim Option ▶

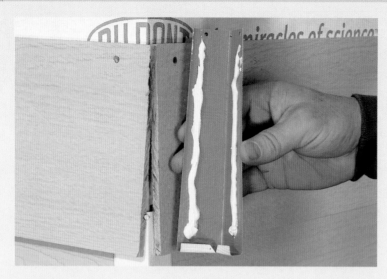

Option: Metal corners made expressly for cement fiber siding eliminate the need for corner trim. Apply caulk to the corner before sliding it in place. Tap the bottom of the corner until it is flush with the bottom edge of the siding. Use a scrap of wood to keep from denting the corner. Blind nail with a 4d galvanized siding nail.

Measure and cut trim to fit around a window. Install trim along the bottom of the window first, then measure and cut trim to fit along the sides, flush with the bottom edge of the first trim piece and ⅛" above the top of the drip cap. Measure and cut trim to fit over the window, flush with the outside edges of the side trim. Drive two nails in the trim pieces every 16". Repeat for each window and door.

On horizontal eaves, install frieze boards directly under the soffits. Butt the frieze boards against the corner trim, and drive two nails every 16" into studs. *Note: Wider frieze boards usually look better if they're run continuously around the house, with corner boards installed afterward.*

7

Install frieze in gable ends. Use a T-bevel to determine the angle on the gable end of the house. Cut this angle on the end of a frieze board, and install under the soffits.

8

Install wood lath along the base of the walls. Align the bottom edge of the lath with the chalkline and nail in place using 6d nails. Keep the lath ⅛" from the corner trim. *Tip: Rather than buying lath, rip panels of wood or fiber-cement siding to 1½"-wide strips, and use them as lath.*

9

Make a story pole as a guide to the location of each course. Use it to avoid slivers of siding above or beneath windows. Starting with the front of your house, work your way around to come up with the best positioning for each course.

Base Trim Alternative ▸

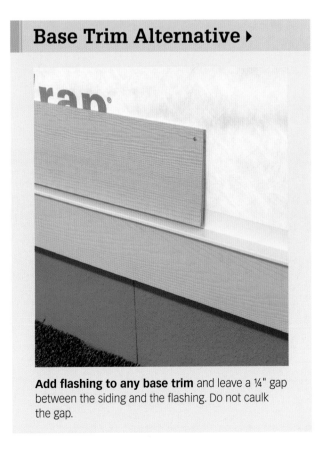

Add flashing to any base trim and leave a ¼" gap between the siding and the flashing. Do not caulk the gap.

(continued)

Cutting and Priming Cement Fiber Siding ▸

For minimal dust and noise use power shears designed for cutting cement fiber siding. Support both ends of your work to avoid binding.

Prime all freshly cut edges to limit moisture damage and extend the life of the siding. A can of quick-drying spray primer is a handy way to get the job done.

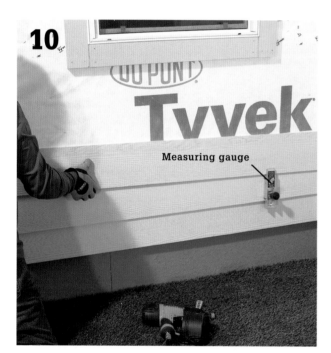

10

Measuring gauge

11

Set the measuring gauge to give the panels a minimum of 1¼" overlap. Place the second row of panels over the first, using the measuring gauge to set the amount of overlap. Offset seams by at least one stud. Repeat this procedure for subsequent rows. Check every five or six rows for level. Make adjustments in small increments. Cut or notch panels as necessary to fit around protrusions in the walls.

Accurate marking along corner trim is made easy with this simple guide. In a scrap of 1×, cut a notch slightly wider than the exposure and ⅛ inch deeper than the thickness of the siding. Push the guide firmly against the corner trim and mark.

12

Cut the first siding panel so it ends halfway over a stud when the other end is placed ⅛" from a corner trim board. Apply primer to the cut end. Hang the siding over the bottom edge of the lath. Keep a ¹⁄₁₆" to ⅛" gap between the siding and corner trim. Nail the panel at each stud location 1" from the top edge using siding nails.

13

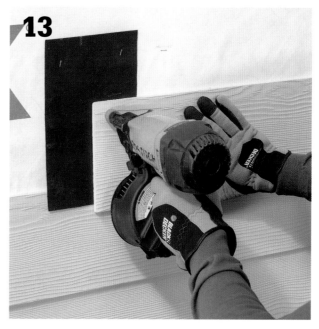

Measure and cut the next panel so it reaches the opposite corner or falls at the midpoint of a different stud. Nail the panel ⅜" from the seam edge and at every stud. Staple felt or metal flashing behind each joint. Do not caulk butt joints.

14

For windows, slide the siding panel against the bottom window trim. Mark the panel ⅛" from the outside edges of the side trim. Place a scrap piece of siding next to the window trim at the proper overlap. Mark the depth of the cut ⅛" below the bottom trim. Transfer the measurement to the siding panel and cut it to fit. Install the cutout panel around the window. Do the same at the top of the window.

(continued)

To cut a notch with power shears, make a curved cut to remove the bulk of the notch. Then reverse your position to complete the cut parallel to the edge of the plank.

To complete a notch make two perpendicular cuts, stopping short of the cut line. Remove any remaining bits in the corners with a utility knife.

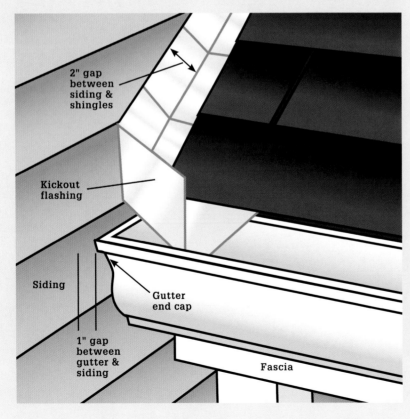

2" gap between siding & shingles

Kickout flashing

Siding

Gutter end cap

1" gap between gutter & siding

Fascia

Add kickout flashing where a roof meets a wall to protect the siding from the water that flows down a sloped roof—good practice for any type of roofing. Fabricate the piece from aluminum or galvanized steel sheet metal. When installing the siding over a roofline, keep the planks 2" above the roofing. Use a T-bevel to determine the angle of the roofline, and transfer the angle to the siding. Avoid nailing into the flashing.

15

Rip the last row of panels to fit below the frieze boards under the horizontal eaves. Nail the panels in place.

16

Use a T-bevel to determine the angle of the roof line on the gable ends of the house. Transfer the angle to the panels, and cut them to fit.

17

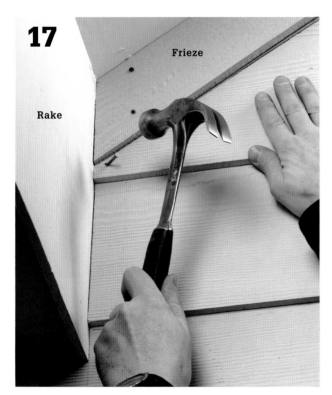

Keep the panels ⅛" from the rake boards, and nail them in place along the gable end.

18

Fill all gaps between panels and trim with flexible, paintable caulk. Paint the siding as desired (see Painting & Staining Siding, beginning on page 200).

Wood Shakes & Shingles

The siding application for shakes and for shingles is basically the same, but there is a difference between the two materials. Shakes are hand split then sawn in half, giving them a rough surface and a flat, smooth back. Shingles are machine sawn on both sides and sanded to create smooth, tapered boards that are thinner than shakes. Both shakes and shingles are typically made from cedar.

Shakes and shingles are usually installed in a pattern called single coursing, which is what's used for our project. Each piece of siding overlaps the one below. For a greater look of depth, consider the double course or staggered butt course shown on the opposite page.

Each shake or shingle is installed with two corrosion-resistant nails, most commonly 4d to 7d nails. The nail size is determined by the size and type of shake or shingle. Check manufacturer's recommendations. The manufacturer will also specify the spacing between siding pieces. Most shakes and shingles will expand after installation, although some that are green or freshly cut may actually shrink.

The siding must be installed over sheathing that has a solid nailing surface, such as plywood. If there isn't a suitable nailing surface, install 1 × 3 or 1 × 4 furring strips across the house for nailing. Felt paper is commonly used as an underlayment, but check building codes for underlayment requirements in your area.

Tools & Materials ▸

Hammer	Caulk gun
Handsaw	Shakes or shingles
or coping saw	Cedar trim boards
Tape measure	(2 × 2, 1 × 3,
Utility knife	1 × 4)
Stapler	Nails
Chalkline	30# felt paper
Line level	Staples
Paintbrush	Sealer
T-bevel	Flexible caulk

Wood shingles are an attractive mainstay of Craftsman homes, as well as Cape Cod and Victorian styles. Featured on a gable or two, shingles and shakes can be an asset for any home.

Variations for Coursing

Double-course installation offers a greater look of depth between rows. Each course of shingles is installed ½" lower than an undercourse that's placed beneath it. A lower-grade, less-expensive shingle is typically used for the undercourse.

Staggered butt–course installation features a random, three-dimensional look. The application starts with a double starter row, and shingles in overlapping courses are staggered by up to 1".

Variations for Corner Installations

Woven corners have shingles that overlap at the corners for a weave effect. This overlap alternates between walls with each successive course.

Mitered corners are made by cutting corner shingles at a 45° angle and butting them together. This method is very time-consuming.

How to Install Single-course Wood Shakes & Shingles

Cover the exterior sheathing with felt paper. Starting at the bottom of a wall, install the paper horizontally using staples. Wrap corners a minimum of 4". Overlap vertical seams 6" and horizontal seams 2". Cut out openings around doors and windows.

Starting at the lowest corner of the house, snap a level chalkline at the bottom of each wall where the siding will start. Measure the height of the wall from the chalkline to the soffits.

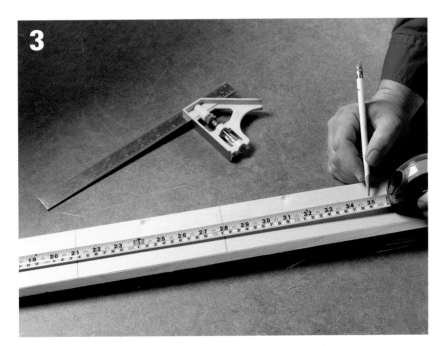

To determine the exposure of the shingles—the amount of wood revealed beneath the overlap—divide the wall height by the number of proposed rows. The goal is to find an exposure measurement that can be multiplied by a whole number to equal the wall height—a 120" wall can have 12 rows with a 10" exposure, for example. Create a story pole on a straight 1 × 3 by making a series of marks equal to the exposure. *Tip: It's best for rows to align with the tops and bottoms of doors and windows.*

Place the story pole at a corner, aligning the bottom with the chalkline. Transfer the marks onto the wall. Do this at each corner, door, and window location.

Place a 1 × 3 board at an outside corner, aligning it with the chalkline at the bottom. Keep the outside edge flush with the adjacent wall. Nail it in place. Overlap the edge of the board with a 1 × 4, align it with the chalkline, and nail in place. If more than one trim board is needed to span the length of the wall, miter the ends at 45° and place them together. Do this for all outside corners. *Tip: As soon as you cut a shingle or piece of trim, apply sealer to the cut edge.*

Fasten a 2 × 2 in an inside corner flush with the chalkline at the bottom. If more than one board is needed to span the height of the wall, miter the adjoining ends at 45° and butt them together.

If the chalkline is hard to see, run a string from the bottom of the corner trim pieces. Starting at a corner, install the starter row of shingles ½" above the chalkline. Keep the manufacturer's recommended distance between shingles and between shingles and trim, usually a ⅛ to ¼" gap. Keep nails ¾" from edges and 1" above the line of exposure.

Place the first course of shingles over the starter row flush with the chalkline at the bottom. Overlap the seams in the starter course by at least 1½".

(continued)

Snap a chalkline across the shingles to mark the exposure using the reference lines from step 4. Install a course of shingles at the chalkline, offsetting the seams. Install remaining rows the same way.

Option: To ensure straight lines, tack a 1 × 4 flush with your reference lines. Nail the board through gaps between shingles. Use the board as a guide for installing the shingles.

Cut shingles to fit around doors, windows, and protrusions in the walls using a coping saw or handsaw. Make sure gaps between shingles aren't aligned with the edges of doors and windows. *Tip: Whenever possible, plan your layout so you can install full shingles next to doors and windows rather than cutting shingles to fit.*

Cut and install the tops of shingles over windows and doors. Align the tops with adjacent shingles on either side of the door or window.

When applying shingles on a dormer or above a roof line, keep the bottom edge of the shingles ½" above the roof shingles. Cut shingles with a handsaw or circular saw.

Measure and cut the last row of shingles to fit under the horizontal eaves. Leave the recommended gap between shingles and soffits. Nail the shingles in place.

Use a T-bevel to determine the roof angle on the gable end of the house. Cut shingles at this angle for each row at the gable end using a handsaw or circular saw. Install shingles up to the peak. *Tip: As you near the top of a wall, measure from the soffits to the shingles on each side of the wall. If the measurements aren't equal, make small adjustments in successive rows until the distance is the same.*

Caulk around all doors, windows, protrusions, and corner trim. Apply your choice of stain or primer and paint (see Painting & Staining Siding, starting on page 200).

Plywood Siding

Often under-rated, plywood and other panel-type siding offer an inexpensive solution that installs quickly and with the right trim treatment can look great. In fact, it is the siding of choice for stylish contemporary homes because of its clean good looks and energy conserving minimal-seam installation.

Horror stories of warping and delaminating are almost always the result of poor installation techniques. Installed with proper flashing, priming, nailing, caulking—and painted regularly— panel siding can last indefinitely.

The size of panels (up to 12 ft. long for plywood) is what makes it such a quick install, but can also make it an awkward material to work with. Measurement and cutting have to be precise. A helper is invaluable. Some of this difficulty is eliminated if you install 5/4 base trim that functions as a handy ledge while installing and provides protection for the often-vulnerable bottom edge of the panel. Alternatively, a temporary ledger lets you slide panels into place.

The steps that follow emphasize plywood panels, but the same techniques apply to pressed hardboard, OSB (oriented strand board) panels, and with the exception of cutting techniques, cement fiber panels. Always check the manufacturer's requirements for gaps between and around panels. Ignoring such recommendations is almost always the cause of buckling and warping.

Trim options include imitation board and batten, grids, mock Tudor, or none at all as in the case of some rain screen approaches. Additionally, designers of contemporary homes often find clever ways of using an inch or two gap between panels as a design feature.

Tools & Materials ▸

Tape measure	Plywood panels
Circular saw	1 × 4 trim
Hammer	2 × 6 temporary ledger,
Chalkline	with stakes
4-foot level	1 × 3 battens (optional)
Drywall T-square	6d, 8d, and 10d
Stud finder	corrosion-resistant
Pry bar	box nails
Tin snips	Drip caps
Jigsaw	Z-flashing
Sanding block	Caulk
Caulk gun	Eye protection
Paintbrush	Gloves

Panel siding's crisp good looks are just one of its benefits. Low-cost, ease of installation, and energy conservation are also appealing traits.

How to Install Plywood Siding

Install a temporary ledger as a handy base for placing panels. A 2 × 6 held by 2 × 4 stakes does the job. Apply housewrap (see pages 136–137) and install flashing above windows and doors. Use a stud finder to mark the location of vertical studs. If the edges of your panels are unprimed, prime them in advance.

Set up a cutting area (see page 129) using sawhorses and four support boards. Working from a corner, measure for your first panel, taking care that its edge will end up centered on a stud for effective nailing. To mark for a rip cut, snap a chalkline or mark along a drywall T-square from both ends of the panel.

Make a cutting guide using a 1 × 4 fastened to a strip of ½" plywood 8 feet long. As you fasten the 1 × 4, align it with the factory edge of the plywood. Next, cut the plywood, running your circular saw along the 1 × 4. By clamping the plywood edge along your cut, you'll have a no-fuss cutting guide.

Cut the first panel wider than needed so you can adjust for any lack of plumbness to a corner. For an inside corner, make the piece 3½" too wide, then use a 1 × 4 to scribe a line while a coworker checks for plumb. Rip the panel, position it, and recheck for plumb. If it looks good, nail with galvanized siding nails long enough to penetrate 1½" into framing. Nail every 6" around the perimeter and every 12" into studs in the field of the panel.

(continued)

Slide the next panel into place, taking advantage of your ledger as a support. Leave a gap in the vertical joint according to manufacturer's instructions—typically ¼". Drive the nails until they are flush with the surface. Don't overdrive the nails so they crack into the veneer. The resulting pockmark will harbor moisture and can lead to rot.

Mark for windows and doors by sliding the panel alongside the obstruction. Allow for the recommended gap at the sides and bottom and over the window and door flashing. Measure from the nearest installed panel to determine the extent of the notch.

To make the cutout for a window or door, use a circular saw for the cuts perpendicular to the long edge of the panel. Make a plunge cut (see page 148) for long cuts parallel to the edge. In both cases saw only up to the cut line. Finish the cuts with a jigsaw.

To make a cross cut, flip the panel and saw with the back side up to limit any splintering on the face of the panel. Start the cut and clamp a framing square or other straight edge in place as a guide. Use a sanding block to tidy up the cut.

9

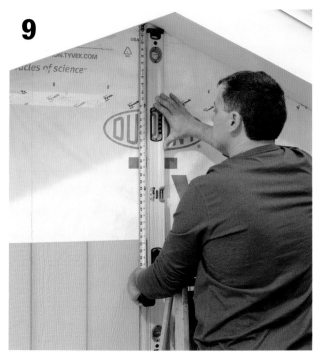

To measure for an angle cut, hold a level plumb where the edge of the panel will reach. Measure alongside the level, allowing for ¼" gap above any flashing. Next, measure from the center of the stud to the starting point of the angle cut—the base of the triangular piece.

10

Spacers help maintain the gap required above flashing. This gap allows moisture to evaporate. Use scraps of ¼" stock as spacers to position the piece while you nail.

11

Leave a large gap above roofing to guard the siding from prolonged contact with water running down the roof. Check manufacturer's recommendations and rip a piece of wood for a spacer to maintain the gap while you nail.

12

Finish with corner trim and battens. When applying corner trim, install the overlapping piece on the side of the house where looks count. Install battens by tacking them at the top and plumbing them with a level. Use 10d galvanized casing nails to fasten trim.

Board & Batten Siding

Board and batten is a vertically installed siding that offers a rustic look and makes houses appear taller. The application consists of installing wide boards vertically on the walls, then placing narrow boards called battens over the seams.

Although there are no set board and batten widths, a popular combination is 1 × 10 boards with 1 × 3 battens, which is what we're using for this project. Our lumber is knotty cedar, which is often used for board and batten applications.

Regardless of the type and size of lumber you choose, the battens need to overlap the boards by at least ½". To maintain a uniform look, measure the length of the wall, and the width of the boards, and determine the appropriate spacing between boards so you can install a full board at the end of the wall. Ripping the last board to fit will ruin the symmetry you've established.

Before installing the siding, you'll need to install horizontal blocking lines or furring strips on the walls. This gives you a firm nailing base for the siding. Nails must be driven 1½" into solid wood. Nailing directly to your sheathing, even if it's plywood, will not provide sufficient hold. Because the nailing strips add thickness to your walls, you'll need to extend the jambs and sills around your doors and windows.

Another way to achieve the board-on-board look is to use board and batten panels (page 121). These panels feature a reverse batten style, in which the batten is placed behind the simulated boards. The panels are typically available in 4 × 8- or 4 × 9-ft. sheets to span the length of the wall. Be sure to purchase panels that are rated for exterior use.

Tools & Materials ▶

Tape measure	4-ft. level
Circular saw	Paintbrush
Hammer	Cedar lumber (1 × 10, 1 × 3)
Chalkline	1 × 3 nailing strips
Line level	2 × 4
Pry bar	8d and 10d corrosion-
Tin snips	resistant box nails
Level	Drip caps
Jigsaw	6d galvanized box nails
T-bevel	Sealer
Caulk gun	Caulk

The board and batten siding on the second level offers vertical, parallel lines that are juxtaposed against the horizontal siding on the lower level. The effect is a contrasting exterior that uses multiple colors.

How to Install Board & Batten Siding

1

Cover the walls with housewrap (pages 136 to 137). Starting at the bottom of the wall, fasten horizontal nailing strips every 16 to 24". Nail the strips to studs using 8d nails. Install nailing strips around all doors and windows. Do this for each wall.

2

Replace the exterior trim with new extenders ripped to the same width as the jamb and thick enough to be flush with the board siding when installed. Using galvanized finishing nails that penetrate the jamb at least 1 inch, install extenders around the perimeter of the window.

3

Cut a piece of drip cap to size to fit over the window using tin snips. Set the drip cap in place so the bottom lip is over the jamb extension. Nail the drip cap in the upper corners using 6d galvanized nails. Repeat steps 2 and 3 for each window and door.

4

Install ledger here

Starting at a corner, snap a level chalkline at the bottom of the wall where you want to start the siding. Make sure it's below the nailing strips. Install a straight 2 × 4 flush with the chalkline to use as a temporary ledger. *Tip: Don't cut all boards to size at the start of a wall. The distance from the ledger to the soffits can change, which can impact the length of the boards.*

(continued)

5

Measure from the ledger to the soffits, then subtract ⅛".
Cut a siding board to this length. Set the bottom of the board
on the ledger and align the side with the edge of the wall. For
boards 6" and narrower, drive one 8d nail in the center at each
nailing strip. For boards wider than 6", drive two nails 3" apart.

6

Wood spacer

Nail "handle"
for spacer
removal

Cut the next board to size and set in place on the ledger.
Keep your predetermined gap between boards, but make
sure the batten will overlap each board by at least ½". Nail in
place at each nailing strip. Install remaining boards the same
way. Check every few boards with a level to make sure they're
plumb. If they're not, adjust slightly until plumb. To maintain
even spacing between boards, cut wood spacers the size of
your gaps and use them when installing each board.

7

At windows, set a board on the ledger and push it against
the window frame. Mark ⅛" above and below the frame. Using
a scrap to maintain the gap, hold a scrap of wood under the
window and mark ⅛" outside the side jamb. Mark and make
the cutout using a jigsaw.

8

Set the board in place, keeping a ⅛" gap around the
window jamb. When installing, drive only one nail per nailing
strip in the area next to the window. Repeat steps 7 and 8 for
each window and door.

9

If more than one board is needed to span the height of a wall, cut a 45° bevel in the ends of the adjoining boards and butt them together. Make sure the seam falls over a nailing strip. Offset seams by at least one nailing strip.

10

At the gable end of the house, use a T-bevel to determine the pitch of the roof. Transfer the angle to the boards, cut to size, and install. The length of each board needs to be measured individually since the distance changes along the gable end.

11

Fill gaps between the boards and soffits, and between the boards and jambs with flexible, paintable caulk.

12

Measure and cut battens to size. Center a batten over each gap between boards. Drive one 10d nail in the battens at each nailing strip. Once the battens are installed, remove the ledger board. Paint the siding (see Painting & Staining Siding, starting on page 200).

Log Cabin Siding

Log cabin siding is an inexpensive way to achieve the rustic look of log homes. Since the siding is not composed of full logs, it uses less wood, making it less expensive than traditional log cabins. But from the outside, it's almost impossible to tell that the house is not an actual log cabin. The log-tail corners complete the authentic appearance. Where the log tails would restrict a pathway or entrance, or create a design problem, vertical log corners can be used.

Log cabin siding is generally available in cedar or pine. For our project, we're using 2 × 8 pine siding, which is considerably less expensive than cedar. The siding is facenailed, but the nails are hardly visible. To speed up installation, rent an air compressor and nail gun. If you choose to hand nail, be sure to use a hammer with a smooth face. A corrugated face could mar the siding.

You can cut the siding with a circular saw, but a compound saw or miter saw works better. It's critical for the cuts to be square since the siding pieces butt against each other. The siding must be nailed to studs, so be sure the stud marks on the housewrap are properly aligned with the studs in the walls.

Tools & Materials ▸

Hammer
Level
Chalkline
Chisel
Tape measure
Drill
4-ft. level
Caulk gun
Jigsaw
Miter saw or sliding compound saw
T-bevel
Paintbrush
Log cabin siding
Cedar or pine (2 × 2s, 2 × 4s, 2 × 6s)
Housewrap
Cap nails
Left- and right- side corners
Hot-dipped galvanized siding nails (12d or 16d)
Polyurethane caulk
Sealer

This house isn't really a log cabin, but you'd never know it from the outside. The log cabin siding provides the charm and realism of an actual log home.

How to Install Log Cabin Siding

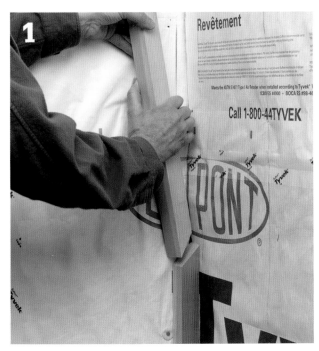

1

Cover the walls with housewrap (pages 136 to 137) and snap a level line around the base of the walls for the first row of siding following step 1 on page 149. Place a 2 × 2 in each inside corner flush with the chalkline at the bottom, and nail in place. If more than one 2 × 2 is needed to span the wall, cut a 45° bevel at the joining ends.

2

2 × 4 casing   ⅛" gap

Use 2 × 2s or 2 × 4s to trim doors and windows.
Measure the top, bottom, and sides of doors and windows, add ¼" to each piece, and cut to size. Install the trim, keeping a ⅛" gap between the window and door frames and the trim.

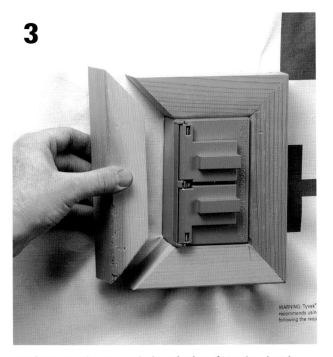

3

Apply 2 × 2 trim around electrical outlets, the electric meter, and vents.

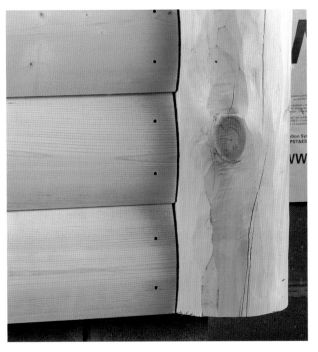

Option: If using vertical corners, place the corner on the wall flush with the chalkline at the bottom, and nail in place. Install vertical corners before applying any siding. Keep siding ⅛" from these corners.

(continued)

4

Hold a piece of siding in place with the bottom (groove) edge aligned with the chalkline (step 1) and the end flush with an outside corner. Mark the opposite end at the midpoint of the last stud it crosses. Cut the siding at the mark. Apply sealer to the cut end. *Tip: Apply sealer to the cut ends of siding, trim, and corners before installing.*

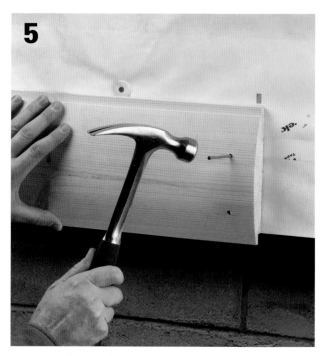

5

Set the siding back in place along the chalkline flush with the corner. Fasten it to the wall with two siding nails at each stud location. Drive the first nail 1½" from the bottom edge, and the second nail 3 to 4" above that. Cut the next piece to reach the opposite corner, butt it against the first piece, and install.

6

On the adjacent wall, start with a corner piece. Hold it in place so the log tail overlaps the adjacent siding. Mark the opposite end at the last stud. Cut it to length, set it along the chalkline so the corner overlaps the adjacent wall, and nail in place. Install the remaining first course the same way.

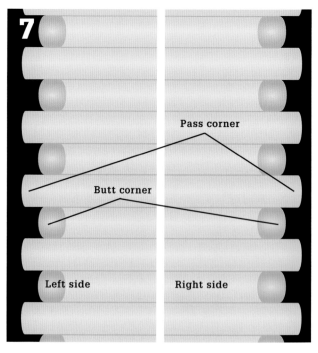

7

Pass corner

Butt corner

Left side   Right side

Alternate between the butt and the pass at the corners for each row of siding. The corners at both ends of a wall must be the same, either both butts or both passes, for each course. You cannot have a butt at one end of the wall and a pass at the opposite corner.

8

Place the second row of siding over the first, setting the groove over the lip of the siding below. Offset joints between rows by at least two studs. Keep a ⅛" gap between the siding and inside corners. Install remaining courses the same way. *Tip: There are left-side and right-side corner pieces. Be sure to use the correct piece for each corner. Start the siding on a wall that has a "butt" corner, then overlap it with a "pass" on the adjacent side. For a butt corner, the end of the siding is flush with the corner. A pass corner extends past the corner to overlap siding on the adjacent wall.*

9

For hose spigots and other small wall protrusions that you don't frame around, drill a hole in the siding at the proper location, then place the siding over the object. To keep the hole as small as possible, you may need to remove the protrusion, then reinsert it after the siding is in place.

10

Check every few rows of siding with a level. If necessary, leave a small gap between the grooves and lips in the siding until the rows are level. Make the changes subtly over several courses. *Tip: Start with 8-ft. sections of log-tail corner pieces for the first row. With each successive course, move the joints over two studs until you can use 4-ft. sections, then start again with 8-ft. pieces.*

(continued)

Install siding up to the bottom of windows. Hold a piece of siding in place below the window framing. Make a mark on the siding ⅛" past the outside edge of the side window trim. Place a scrap of siding next to the window over the last installed row. Mark the siding piece ⅛" below the edge of the window.

Transfer the measurements from the last step to a piece of siding. Cut out the opening using a jigsaw.

Install the siding, keeping a ⅛" gap around the window frame. As you continue installing rows of siding, maintain a ⅛" gap between the siding and trim.

Follow steps 11 to 12 to mark and cut siding to fit over the top of doors and windows. Center the siding opening over the door and window, and nail in place. *Tip: For best appearance, keep joints in the siding from falling directly above or below windows and doors.*

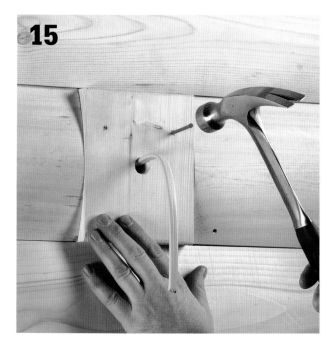

15

At light fixture locations, drill a hole in the siding for electrical wires. To make a flat surface to hold the fixture, start at the wire hole and work out 2¾" on each side, making a series of 1"-deep cuts in the siding. Chisel out the wood until the surface is flat, and apply sealer. Install the siding, feeding the wires through the hole. Cut a 2 × 6 the height of the siding, drill a hole for the wiring, and install in the notch. Mount the fixture on the 2 × 6.

16

For the top row, measure the distance from the bottom of the lip of the last installed row of siding to the eaves. Subtract ⅛" and rip siding to this measurement. Nail the siding in place under the eaves.

17

On the gable ends of the house, use a T-bevel to determine the roof angle. Cut the ends of the siding at this angle, then install along the gable.

18

Caulk between the siding and inside corners. Also caulk between siding and vertical corners, siding and window and door framing, and around wall protrusions. Do not apply caulk to the joints between siding. Stain the siding following the instructions that start on page 200.

Brickmold

Brickmold is decorative trim that forms a transition between siding and windows or doors. It's usually made of pine or fir. Unless these moldings are maintained regularly with fresh coats of paint, they are prone to rotting. Brickmold that isn't primed on both the front and back surfaces is particularly vulnerable to rot. If you are replacing your home's siding, have a close look at your current brickmold. Soft spots in the wood are telltale signs of rotting behind the paint. Rotting also occurs where the mitered molding meets in the corners. If you find rotted areas, replace the affected molding. You can remove and replace brickmold fairly easily and at any time, without removing siding.

Tools & Materials ▶

Pry bar	Caulk gun
Measuring tape	Brickmold
Pencil	Drip edge
Combination square	10d exterior
Miter saw	casing nails
Hammer and nail set	Caulk

How to Replace Damaged Brickmold

Remove old sections of brickmold with a pry bar. You may find it helpful to first insert a stiff-bladed paint scraper between the brickmold and underlying jamb to break the paint seal.

The best time to paint your new brickmold is before installing it. Prime all surfaces of the molding if it doesn't come preprimed, and paint the outer surfaces your trim color. Paint all your brickmold stock now in long strips.

Install drip edge. If no metal drip edge was in place above the old brickmold, cut a strip to length and slip it up into place behind the siding. Drip edge will extend the life of wood brickmold by keeping water from seeping behind it or soaking the ends.

Measure and cut strips of new brickmold to fit around the window or door. Miter-cut the top ends of the side pieces and both ends of the top piece at 45° using a miter saw. *Note: Some carpenters cut, fit, and nail one strip of brickmold at a time before miter-cutting the next piece. This way, you can adjust the miter angle slightly, if necessary, to improve the fit.*

Install the brickmold strips with 10d exterior casing nails driven every 12" into framing members. Use a nail set to drive the nail heads below the surface of the wood.

Seal the brickmold. Make sure the drip edge is tight against the top brickmold, then apply clear paintable caulk along the top of the drip edge and along the outside edge of the side brickmold where it meets the siding. Fill the nail holes with caulk, and touch up these spots with paint.

Synthetic Brickmold ▸

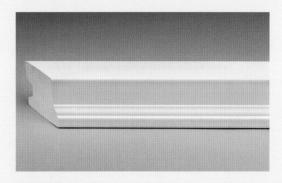

These days, pine and fir aren't the only material options for brickmold and other exterior trim. You can also buy brickmold made of PVC or composites of resin, fiberglass, and wood pulp. Synthetic brickmold is about twice the price of wood brickmold, but it will last indefinitely, it is paintable, and it will not wick or hold water, which could rot surrounding framing. If you're willing to pay more for synthetics, the tradeoffs are worth the extra cost: you'll never have to maintain or replace your brickmold again.

Finishing Walls with Masonry

The thought of hefting heavy rock or mixing batch after batch of stucco mix may sound like miserable work for those who have never tried a masonry project, but in reality masonry isn't as difficult as it may seem. In fact, all it may take is a dash of brick or veneered stone here and there to give your home the charm that masonry can bring, without much work at all. Granted, coating every wall with stucco is a big job that's better left to those who make a living doing it, but don't be afraid to try your hand at it. With a small collection of

masonry tools and maybe a rented mixer, you'll have what you need to tackle the job yourself. This chapter will walk you through the process of completing five different masonry projects. One of them requires no mortar at all.

As with any roofing or siding project, make sure to correctly prepare walls before you begin the job. Proper wrapping, flashing, and structural reinforcement will impact how long and well your masonry project stands the test of time.

Masonry Tools & Materials

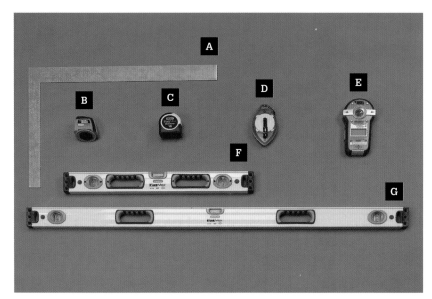

Layout and measuring tools for installing and leveling include: framing square (A), stud finder (B), tape measure (C), chalkline (D), laser level (E), 2-foot level (F), and 4-foot level (G).

Mixing and pouring tools include masonry hoe and mortar box for mixing small amounts of concrete; garden hose and bucket for delivering and measuring water; and power mixer for mixing medium-sized loads of concrete.

Using the right tools for your masonry projects make the job much easier and will save a lot of time. The most common tools for masonry siding projects such as brick, stone, and stucco are as follows: mortar bag (A) for applying clean lines of mortar and grout; inside corner tool (B) for shaping and siding inside corners of concrete; outside corner tool (C) for shaping and forming outside corners of concrete; whisk broom (D) for applying different finishes and textures to concrete; mortar hawk (E), also known as mortarboard, for holding mortar while laying brick, block, or stone; wood flat (F) for flat application of latex or stucco; jointing tools (G, H, I, J), also known as jointers, for forming joints in brick, block, or stone are available in different sizes and shapes to match your specific joint width; wide-mouth nippers (K) for cutting stone; aviation snips (L) for cutting stucco lath; groover (M) for forming joints in concrete slabs or treads in steps; edger (N) for shaping and forming concrete edges; mason's hammer (O) for setting or cutting brick, block, or stone; mason's trowel (P) for mixing and moving mortar; hand maul (Q) for breaking up concrete and tapping bricks or stone into place in mortar; square-end trowel (R) for pressing mortar into lath.

Brick

Brick veneer is essentially a brick wall built around the exterior walls of a house. It's attached to the house with metal wall ties and supported by a metal shelf hanger on the foundation. It's best to use queen-sized bricks for veneer projects because they're thinner than standard construction bricks. This means less weight for the house walls to support. Even so, brick veneer is quite heavy. Ask your local building inspector about building code rules that apply to your project. In the project shown here, brick veneer is installed over the foundation walls and side walls up to the bottom of the windowsills on the first floor of the house. The siding materials in these areas are removed before installing the brick.

Construct a story pole before you start laying the brick so you can check your work as you go along to be sure your mortar joints are of a consistent thickness. A standard ⅜" gap is used in the project shown here.

Tools & Materials ▶

Hammer
Circular saw
Combination square
Level
Drill with masonry bit
Socket wrench set
Staple gun
Mason's trowel
Masonry hoe
Mortar box
Mason's chisel
Maul
Pressure-treated 2 × 4s
⅜ × 4" lag screws and washers
2 × 2
Lead sleeve anchors
Angle iron for metal shelf supports
30 mil PVC roll flashing
Corrugated metal wall ties
Brickmold for sill extensions
Sill-nosing trim
Type N mortar
Bricks
⅜-dia. cotton rope

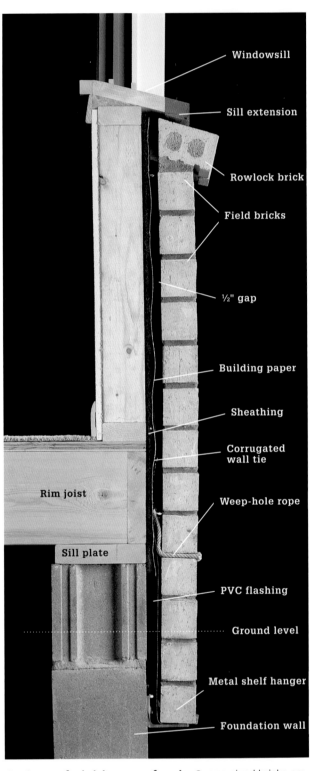

Anatomy of a brick veneer facade: Queen-sized bricks are stacked onto a metal or concrete shelf and connected to the foundation and walls with metal ties. Rowlock bricks are cut to follow the slope of the windowsills, then laid on edge over the top course of bricks.

Image labels: Windowsill · Sill extension · Rowlock brick · Field bricks · ½" gap · Building paper · Sheathing · Corrugated wall tie · Weep-hole rope · PVC flashing · Ground level · Metal shelf hanger · Foundation wall · Rim joist · Sill plate

How to Install Brick Veneer

1

Sill extension

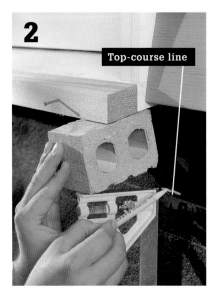

2

Top-course line

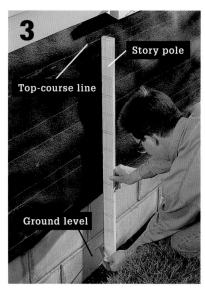

3

Top-course line

Story pole

Ground level

Remove all siding materials in the area you plan to finish with brick veneer. Before laying out the project, cut the sill extension from a pressure-treated 2 × 4. Tack the extension to the sill temporarily.

Precut the bricks to follow the slope of the sill and overhang the field brick by 2". Position this rowlock brick directly under the sill extension. Use a combination square or level to transfer the lowest point on the brick onto the sheathing (marking the height for the top course of brick in the field). Use a level to extend the line. Remove the sill extensions.

Make a story pole long enough to span the project area. Mark the pole with ⅜" joints between bricks. Dig a 12"-wide, 12"-deep trench next to the wall. Position the pole so the top-course line on the sheathing aligns with a top mark for a brick on the pole. Mark a line for the first course on the wall below ground level.

4

5

Web locations

Masonry anchor

Extend the mark for the first-course height across the foundation wall using a level as a guide. Measure the thickness of the metal shelf (usually ¼") and drill pilot holes for 10d nails into the foundation at 16" intervals along the first-course line, far enough below the line to allow for the thickness of the shelf. Slip nails into the pilot holes to create temporary support for the shelf.

Set the metal shelf onto the temporary supports. Mark the location of the center web of each block onto the vertical face of the shelf. Remove the shelf and drill ⅜"-dia. holes for lag screws at the web marks. Set the shelf back onto the temporary supports and outline the predrilled holes on the blocks. Remove the shelf and drill holes for the masonry anchors into the foundation using a masonry bit. Drive masonry anchors into the holes.

(continued)

Reposition the shelf on the supports so the predrilled holes align with the masonry anchors. Attach the shelf to the foundation wall with 3/8 × 4" lag screws and washers. Allow 1/16" for an expansion joint between shelf sections. Remove the temporary support nails.

After all sections of the metal shelf are attached, staple 30 mil PVC flashing above the foundation wall so it overlaps the metal shelf.

Test-fit the first course on the shelf. Work in from the ends using spacers to set the gaps between bricks. You may need to cut the final brick for the course. Or, choose a pattern such as running bond that uses cut bricks.

Build up the corners two courses above ground level, then attach line blocks and mason's string to the end bricks. Fill in the field bricks so they align with the strings. Every 30 minutes, smooth mortar joints that are firm.

Attach another course of PVC flashing to the wall so it covers the top course of bricks, then staple building paper to the wall so it overlaps the top edge of the PVC flashing by at least 12". Mark wall-stud locations on the building paper.

Use the story pole to mark layout lines for the tops of every fifth course of bricks. Attach corrugated metal wall ties to the sheathing where the brick lines meet the marked wall-stud locations.

Fill in the next course of bricks, applying mortar directly onto the PVC flashing. At every third mortar joint in this course, tack a 10" piece of ⅜"-dia. cotton rope to the sheathing so it extends all the way through the bottom of the joint, creating a weep hole for drainage. Embed the metal wall ties in the mortar beds applied to this course.

Add courses of bricks, building up corners first, then filling in the field. Embed the wall ties into the mortar beds as you reach them. Use corner blocks and a mason's string to verify the alignment, and check frequently with a 4-ft. level to make sure the veneer is plumb.

Apply a ½"-thick mortar bed to the top course and begin laying the rowlock bricks with the cut ends against the wall. Apply a layer of mortar to the bottom of each rowlock brick, then press the brick up against the sheathing with the top edge following the slope of the windowsills.

Finish-nail the sill extensions (step 1, page 181) to the windowsills. Add metal flashing and sill-nosing trim to the siding to cover any gaps above the rowlock course. Fill cores of exposed rowlock blocks with mortar, and caulk any gaps around the veneer with silicone caulk.

Mortarless Brick Veneer

An interesting new siding product is now available that mimics the appearance and durability of classic brick but installs as easily as any other siding material. Mortarless brick veneer systems use stackable bricks to create an appealing façade on wood, steel, or concrete structures. The high-strength concrete bricks are long-lasting—manufacturers offer warranties up to 50 years. And because brick veneer does not require mortar, installation is well within the capabilities of interested homeowners.

Veneer bricks are available in 3" and 4" heights and are either 8 or 9" long, depending on the producer. Bricks weigh approximately 5 pounds each and add 3¼" to the face of walls. While veneer brick systems can be used in both new construction and remodel projects, application is restricted due to the added load: up to 30 feet high on standard wood-framed walls. Consult with a professional builder or structural engineer for walls taller than 30 feet, as well as sections of wall above roofs.

Prior to installation, make sure the framing and wall substrate is sound and the house adequately insulated. Extend all plumbing and electrical pipes, boxes, and meters to accommodate the additional thickness created by the veneer brick and furring strips.

The following pages discuss the installation of brick veneer siding on standard wood-framed walls. All openings require extra support in the form of ¾" plywood lintels. Lintel size is determined by the width of the opening and the brick installation method over the opening (soldier coursing is shown here). Contact the manufacturer or product producer for information regarding lintel sizing, as well as installation of veneer brick on other framing styles.

Tools & Materials ▸

Tape measure	¾" plywood
Chalkline	Furring strips
4-ft. level	Flashing
Utility knife	Self-adhesive
Circular saw	waterproof
Miter saw or wet	membrane (1 × 3,
masonry saw	1 × 4, 1 × 6)
with diamond	Scrap 2 × 4
masonry-cutting	Corrosion-resistant
blade	wood screws
Hammer drill with	(#10 × 2½",
masonry bits	#10 × 4")
Cordless drill with	Outside corner strips
various drivers	Starter strips
Rubber mallet	Veneer bricks
Caulk gun	Outside corner blocks
Work gloves	Inside corner blocks
Safety glasses	Windowsill block
Dust mask	Construction adhesive
Earplugs	Exterior-grade caulk

Brick veneer siding attaches to your house with mechanical fasteners, so you can achieve the appeal of brick without the mess of mortar.

Tips for Installing Brick Veneer Siding

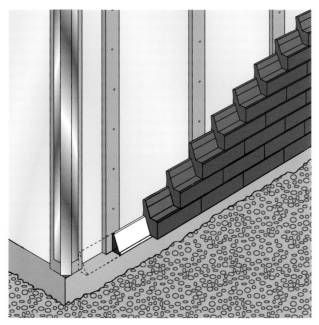

Brick veneer is stacked in courses with staggered joints, much like traditional brick. However, the first course of the mortarless system is installed on a starter strip and fastened with corrosion-resistant screws to 1 × 3 furring strips at each stud location. Bricks are then fastened every fourth course thereafter. At outside corners, a specialty strip is fastened to 1 × 4 furring strips. Corner blocks for both outside and inside corners are secured with screws and construction adhesive.

Before installing brick veneer siding, make sure all openings are properly sealed. For best results, use a self-adhesive waterproof membrane around door and windows. Install the bottom strip first, then the side strips so they overlap the bottom strip. Place the top strip to overlap the sides. Install drip edge flashing where appropriate.

Cut veneer brick using a miter saw with a diamond blade or a wet masonry saw. When cutting brick, protect yourself with heavy work gloves, safety glasses, earplugs, and a dust mask.

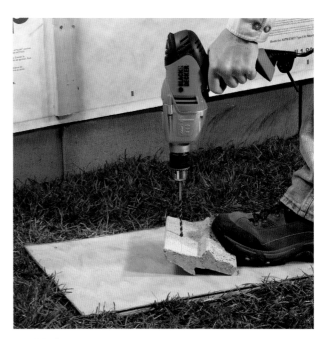

Predrill holes in bricks that require fastening using a hammer drill with a 3/16" masonry bit. Position the brick face-up on the ground and secure with your foot. Drill through the notch in the top portion of the brick, holding the drill bit at 90° to the ground.

How to Install Mortarless Brick Veneer Siding

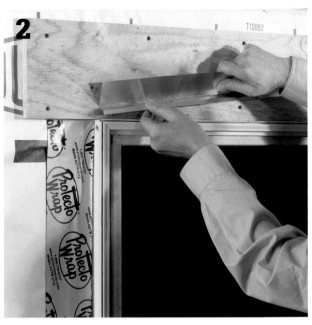

Snap a level chalkline ¾" above the foundation on each wall of the house. Align the bottom end of furring strips above the chalkline. Fasten 1 × 3 furring strips at each stud location with #10 × 2½" corrosion-resistant wood screws. Install 1 × 4 furring strips at outside corners and 1 × 6s at inside corners.

For each opening, cut ¾" plywood lintels to a size of 15" high × 12" longer than the width of the opening. Center the lintel above the opening so 6" extends beyond each side of the frame, and fasten to framing with #10 screws. Install an aluminum drip edge above the window frame, then wrap the lintel and flashing with a strip of self-adhesive waterproof membrane.

At outside corners, position the first section of corner strip 2" above the chalkline. Plumb the strip using a 4-ft. level, then fasten to the framing with #10 × 4" screws every 10" on alternate sides.

Position the starter strip at the chalkline with the flange beneath the ends of the furring strips. Do not overlap corner strips. At inside corners, cut back the starter strip to so it falls 3½" short of adjacent walls. Level the strip, then secure to the framing with #10 × 4" corrosion-resistant wood screws at each furring location.

At the notch in each corner block, predrill a hole at a 30° angle using a hammer drill and a ³⁄₁₆" masonry bit. Place a veneer brick on the starter strip for reference, then slide the first corner block down the corner strip and position it so the bottom edge falls ½" below the bottom edge of the brick. Fasten the brick to the strip with #10 × 2½" wood screws.

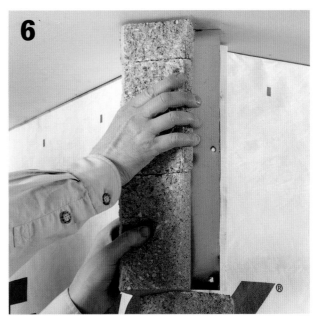

Continue to install corner blocks using #10 × 2½" screws and construction adhesive between courses. For the top of the corner, measure the remaining length and cut a piece of corner strip to size. Fasten blocks to this loose length, cutting the final piece to size if necessary (see page 185). Secure one last block to the existing corner using construction adhesive, then fit the new assembly in place and fasten with #10 × 2½" screws.

For inside corners, predrill holes at 30° angles into inside corner blocks. As with outside corners, position the first block so the bottom edge is ½" below the bottom edge of the first course of veneer brick. Fasten the block to the framing with #10 × 4" wood screws. Continue installing blocks with #10 screws and construction adhesive between each course.

To create the best overall appearance, place a row of bricks on the starter strip so they extend past the width of the most prominent opening on each wall. Place a brick on the second course at each end of the opening, so each sits evenly above the joint of two bricks below. Sight down from the edges of the opening's frame and adjust the entire row to find a pattern that yields the least amount of small pieces of brick around the opening.

(continued)

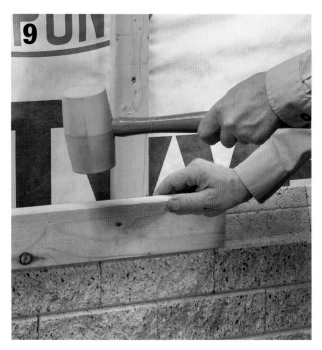

Predrill holes through veneer bricks for the first course (see page 185). Following the established pattern, install bricks on starter strip. At corners, cut bricks to size (see page 185) so they fit snugly against the blocks. Set bricks using a scrap 2 × 4 and rubber mallet to help maintain consistent course alignment.

At each furring strip, hold bricks flat against the wall and secure to the framing with #10 × 2½" screws. Drive screws until the head touches the brick. Do not overtighten.

Fill the brick courses using bricks from different pallets to blend slight variance in color. Set bricks using a scrap of 2 × 4 and a rubber mallet. Check every fourth course for level before fastening bricks to the framing at each furring strip.

To install sill blocks below the widow, fasten a horizontal 1 × 3 furring strip under the window frame extending ⅛" longer than the cumulative width of the sill blocks. Install bricks up to the top of the furring strip, cutting to fit as needed, and fasten each with two #10 × 2½" wood screws. Apply construction adhesive along the top of the furring strip and bricks.

Install the sills, angling them downward slightly, and secure with #10 × 4" wood screws, toenailing through the ends or bottom of the sill into the framing. Cut brick filler pieces to bridge the gap between the sill and the last full course of brick; make sure the pieces align with the rest of the course. Install the pieces with construction adhesive. Seal the gap between the window frame and the sill with exterior-grade caulk.

Continue installing brick along the openings to a height no more than the width of one brick. Cut a piece of starter strip to length, align it with the courses on either side of the opening, and secure to the framing with #10 × 4" screws. Install a course of bricks on the starter strip, fastening them with #10 screws.

Cut bricks for the soldier course to length, then install vertically with two #10 × 2½" screws each. For the final brick, cut off the top portion and secure in place with construction adhesive. For a more symmetrical look, place cut bricks in the center of the course.

At the tops of walls, install 1 × 3 horizontal furring strips. Secure the second to the last course of bricks to the framing with #10 × 2½" screws, then install the last course with construction adhesive. Notch bricks to fit around joists or cut at an angle for gable walls.

Cast Veneer Stone

Cast veneer stones are thin synthetic masonry units that are applied to building walls to imitate the appearance of natural stone veneer. They come in random shapes, sizes, and colors, but they are scaled to fit together neatly without looking unnaturally uniform. Corner stones and horizontal sill blocks allow for easy finishing work.

Cast veneer stones come in a wide variety of sizes and shapes, and are relatively easy to work with and install. Pictured are corner pieces (A), field pieces (B), and a sill block (C).

Tools & Materials ▸

Veneer stone	Stiff-bristle	Jointing tool
Stucco lath	brush	Sill blocks
Type N	Angle grinder	Galvenized L
mortar	w/diamond	brackets
Masonry sand	blade	Metal flashing
Water	Trowel	Safety goggles
	Grout bag	

How to Install Cast Veneer Stone

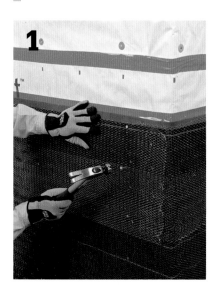

Prepare the wall. Veneer stones can be applied to a full wall or as an accent on the lower portion of a wall. A top height of 36 to 42" looks good. A layer of expanded metal lath (stucco lath) is attached over a substrate of building paper.

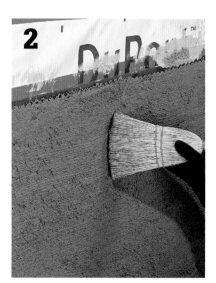

Apply a scratch coat. The wall in the installation area should be covered with a ½- to ¾"-thick layer of mortar. Mix one part Type N mortar to two parts masonry sand and enough water to make the consistency workable. Apply with a trowel, let the mortar dry for 30 minutes. Brush the surface with a stiff-bristle brush.

Test layouts. Uncrate large groups of stones and dry-lay them on the ground to find units that blend well together in shape as well as in color. This will save an enormous amount of time as you install the stones.

Cut veneer stones, if necessary, by scoring with an angle grinder and diamond blade along a cutting line. Rap the waste side of the cut near the scored line with a mason's hammer or a maul. The stone should fracture along the line. Try to keep the cut edge out of view as much as you can.

Apply the stones. Mix mortar in the same ratios as in step 2, but instead of applying it to the wall, apply it to the backs of the stones with a trowel. A ½"-thick layer is about right. Press the mortared stones against the wall in their position. Hold them for a few second so they adhere.

Fill the gaps between stones with mortar once all of the stones are installed and the mortar has had time to dry. Fill a grout bag (sold at concrete supply stores) with mortar mixture and squeeze it into the gaps. Once the mortar sets up, strike it smooth with a jointing tool.

L-brackets

Install sill blocks. These are heavier and wider than the veneer block so they require some reinforcement. Attach three 2 × 2" galvanized L-brackets to the wall for each piece of sill block. Butter the backs of the sill blocks with mortar and press them in place, resting on the L-brackets. Install metal flashing first for extra protection against water penetration.

Stucco

Prized for its weather resistance, durability, and timeless beauty, stucco has long been one of the most popular exterior wall finishes. As a building material, stucco is essentially an exterior plaster made of Portland cement, sand, and water. Other ingredients may include lime, masonry cement, and various special additives for enhancing properties like crack resistance, workability, and strength. With a few exceptions, stucco is applied much as it has been for centuries—a wet mix is troweled onto the wall in successive layers, with the final coat providing the finished color and any decorative surface texture desired.

The two traditional stucco systems are the three-coat system used for standard wood-framed walls, and the two-coat system used for masonry walls, like brick, poured concrete, and concrete block. And today, there's a third process—the one-coat system—which allows you to finish standard framed walls with a single layer of stucco, saving you money and considerable time and labor over traditional three-coat applications. Each of these systems is described in detail on the next page.

The following pages show you an overview of the materials and basic techniques for finishing a wall with stucco. While cladding an entire house or addition is a job for professional masons, smaller projects and repair work can be much more doable for the less experienced. Fortunately, all the stucco materials you need are available in premixed form, so you can be sure of getting the right blend of ingredients for each application (see page 194). During your planning, consult with the local building department to learn about requirements for surface preparation, fire ratings for walls, control joints, and other critical factors.

Tools & Materials ▸

Aviation snips	Square-end trowel
Stapler	Darby or screed board
Hammer	Grade D building paper
Level	Heavy-duty staples
Cement mixer	1½" galvanized roofing nails
Wheelbarrow	Self-furring galvanized
Mortar hawk	metal lath (min. 2.5 lb.)
Raking tool	Metal stucco edging
Wood float	Stucco mix (see page 194)
Texturing tools	Non-sag polyurethane sealant
Flashing	

Stucco is one of the most durable and low-maintenance wall finishes available, but it requires getting each stage of the installation right, as well as the mix of the stucco itself. For this reason, DIYers may want to limit their stucco work to small structures or to repair work only.

Stucco Systems ▸

Three-coat stucco is the traditional application for stud-framed walls covered with plywood, oriented strand board (OSB), or rigid foam insulation sheathing. It starts with two layers of Grade D building paper for a moisture barrier. The wall is then covered with self-furring, expanded metal lath fastened to the framing with galvanized nails.

The first layer of stucco, called the scratch coat, is pressed into the lath, then smoothed to a flat layer about ⅜" thick. While still wet, the stucco is "scratched" with a raking tool to create keys or "tooth" for the next layer to adhere to.

The brown coat is the next layer. It's about ⅜" thick and brings the wall surface to within ⅛ to ¼" of the finished thickness. Imperfections here can easily telegraph through the thin final coat, so the surface must be smooth and flat. To provide tooth for the final layer, the brown coat is finished with a wood float for a slightly roughened texture.

The finish coat completes the treatment, bringing the surface flush with the stucco trim pieces and providing the color and decorative texture, if desired. There are many options for texturing stucco; a few of the classic ones are shown on page 197.

Two-coat stucco is the standard treatment for masonry walls. This system is the same as a three-coat treatment but without a scratch coat. The base coat on a two-coat system is the same as the brown coat on a three-coat system. For the base coat to bond well, the masonry surface must be clean, unpainted, and sufficiently porous. You can test this by spraying water onto the surface: if the water beads and runs down the wall, you should apply bonding adhesive before applying the base coat or you can fasten self-furring metal lath directly to the wall, then apply a full three-coat stucco treatment.

One-coat stucco is a single-layer system for finishing framed walls prepared with a waterproof barrier and metal lath (as with a three-coat system). This treatment calls for one-coat, fiberglass-reinforced stucco, a special formulation that contains alkali-resistant fiberglass fiber, and other additives to combine high-performance characteristics with greatly simplified application. This stucco is applied in a ⅜ - to ⅝"-thick layer using standard techniques. QUIKRETE One Coat Fiberglass Reinforced Stucco meets code requirements for a one hour firewall over wood and form systems.

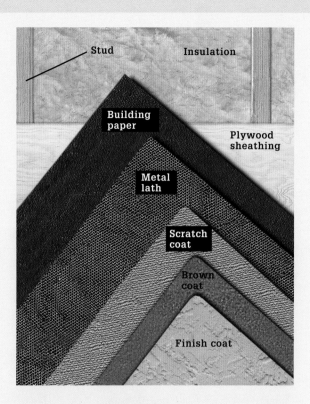

Stud   Insulation
Building paper
Plywood sheathing
Metal lath
Scratch coat
Brown coat
Finish coat

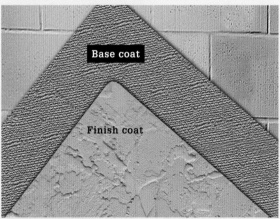

Base coat
Finish coat

Finding the right blend of ingredients and mixing to the proper consistency are critical to the success of any stucco project. Premixed stucco eliminates the guesswork by giving you the perfect blend in each bag, along with mixing and curing instructions for a professional-quality job. All of the stucco products shown here are sold in complete form, meaning all you do is add water before application. Be sure to follow the mixing and curing instructions carefully for each product.

Scratch & Brown, Base Coat stucco: Use this premixed stucco for both the scratch and brown coats of a three-coat application or for the base coat of a two-coat system. You can apply the mixed stucco with a trowel or an approved sprayer. Available in 80 lb. bags in gray color. Each bag yields approximately 0.83 cu. ft. or an applied coverage of approximately 27 sq. ft. at ⅜" thickness.

Finish Coat stucco: Use this stucco for the finish coat on both three-coat and two-coat systems. You can also use it to create a decorative textured finish over one-coat stucco. Apply Finish Coat stucco to a minimum thickness of ⅛", then texture the surface as desired. Available in gray and white for achieving a full range of colors (see below). Coverage of 80-lb. bag is approximately 70 sq. ft. at ⅛" thickness.

One Coat Fiberglass Reinforced Stucco: Complete your stucco application in one step with this convenient all-in-one stucco mix. It's available in gray and white for creating a wide range of colors (see below). You can texture the surface of the single layer or add a top coat of Finish Coat stucco for special decorative effects. Available in 80-lb. bags. An 80-lb. bag covers approximately 25 sq. ft. of wall at ⅜" thickness.

Stucco & Mortar Color: Available in 20 standard colors, Stucco & Mortar Color is a permanent liquid colorant that you blend with the stucco mix before application. Some colors are for use with gray stucco mix, while many others are compatible with white mix. For best results, combine the liquid colorant with the mixing water before adding the dry stucco mix, then blend thoroughly until the color is uniform.

How to Prepare Framed Walls for Stucco

1

Attach building paper over exterior wall sheathing using heavy-duty staples or roofing nails. Overlap sheets by 4". Often, two layers of paper are required or recommended; consult your local building department for code requirements in your area.

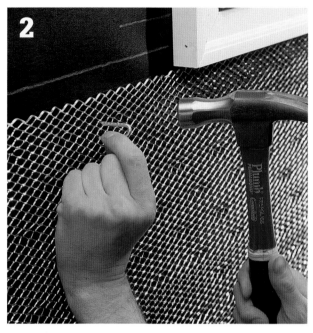

2

Install self-furring expanded metal lath over the building paper with 1½" galvanized fence staples (don't use aluminum nails) driven into the wall studs every 6". Overlap sheets of lath by 1" on horizontal seams and 2" on vertical seams. Install the lath with the rougher side facing out.

3

Install metal edging for clean, finished lines at vertical edges of walls. Install casing bead along the top of stuccoed areas and weep screed (or drip screed) along the bottom edges, as applicable. Make sure all edging is level and plumb, and fasten it with galvanized roofing nails. Add flashing as needed over windows and doors.

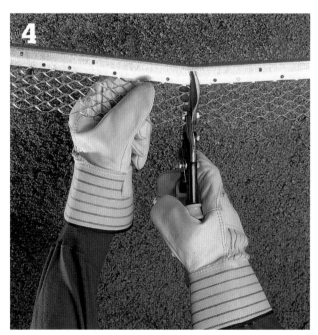

4

Use aviation snips to trim sheets of lath or cut edging materials to length. Cut lath and edging can be very sharp, so always wear gloves when working with these materials.

How to Finish Walls with Stucco

For a three-coat system, mix the stucco to a trowelable consistency and apply it with a square trowel, working from the bottom up. Press the stucco into the lath, then screed and smooth the surface for a uniform thickness.

To trowel a neat outside corner hold a scrap of 1 × 4 tightly against the corner and trowel the stucco up to it. Wait for the stucco to firm up before forming the other side of the corner.

When the coat hardens enough to hold a finger impression, scratch ⅛"-deep horizontal grooves into the surface with a raking tool. You can make a raking tool by drilling a row of holes in a 1 × 2. Use a ⅛" bit and space the holes 1" apart. Insert 2" deck screws or 8d galvanized nails. Let dry per manufacturer's instructions and apply the finish coat in a ⅛" layer (minimum) from the bottom up, covering all walls evenly for color consistency.

Variation: For a one-coat application, mix the stucco and apply it in a ⅜"-thick layer, working from the bottom up and forcing it in to completely embed the lath. Screed the surface flat with a darby or board. When the surface loses its sheen, finish-trowel or texture the surface as desired. Cure the coat as directed. Seal all joints around building elements with polyurethane sealant.

How to Finish Stucco

Test the coloring of finish stucco by adding different proportions of colorant and mix. Let the samples dry to see their true finished color. For the application batches, be sure to use the same proportions when mixing each batch.

Mix the finish batch so it contains slightly more water than the scratch and brown coats. The mix should still stay on the mortar hawk without running.

Finish Option: Cover a float with carpet to make an ideal tool for achieving a float-finish texture. Experiment on a small area.

Finish Option: Achieve a wet-dash finish by flinging, or dashing, stucco onto the surface. Let the stucco cure undisturbed.

Finish Option: For a dash-trowel texture, dash the surface with stucco using a whisk broom (left), then flatten the stucco by troweling over it.

Surface-bonding Cement

Surface bonding cement is a stucco-like compound that you can use to dress up your concrete and cinder block walls. It also adds strength, durability, and water resistance to the walls.

What distinguishes surface bonding cement from stucco is the addition of fiberglass to the blend of Portland cement and sand. The dry mixture is combined with water and acrylic fortifier to form a cement plaster that can bond with concrete, brick, or block for an attractive, water-resistant coating.

Before applying the bonding cement, make sure you have a very clean surface with no crumbling masonry, so the coating can form a durable bond. Because surface bonding cement dries quickly, it's important to mist the brick or block with water before applying the cement so the cement dries slowly. As with most masonry projects, the need to dampen the wall increases in very dry weather.

The bonding cement can be used on both mortared and mortarless walls, and on both load-bearing and non-load-bearing walls. However, it is not recommended for walls higher than 15 courses of block.

Tools & Materials ▸

Garden hose with
    spray attachment
Bucket
Wheelbarrow
Mortar hawk
Square-end trowel
Groover
Caulk gun
Surface
    bonding cement
Concrete
    acrylic fortifier
Tint (optional)
Silicone caulk

Mix small batches of dry surface bonding cement, water, and concrete acrylic fortifier according to the manufacturer's instructions until you get a feel for how much coating you can apply before it hardens. An accelerant in the cement causes the mix to harden quickly—within 30 to 90 minutes, depending on weather conditions. The cement can be tinted before application.

How to Finish Walls with Surface Bonding Cement

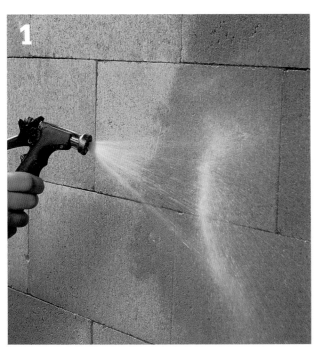

1

Starting near the top of the wall, mist a 2 × 5-ft. section on one side of the wall with water to prevent the blocks from absorbing moisture from the cement once the coating is applied.

2

Mix the cement in small batches according to the manufacturer's instructions, and apply a ¹⁄₁₆- to ¹⁄₈"-thick layer to the damp blocks using a square-end trowel. Spread the cement evenly by angling the trowel slightly and making broad upward strokes.

3

Use a wet trowel to smooth the surface and create the texture of your choice. Rinse the trowel frequently to keep it clean and wet.

4

To prevent random cracking, use a groover to cut control joints from the top to the bottom of the wall every 4 ft. for a 2-ft.-high wall, and every 8 ft. for a 4-ft.-high wall. Seal hardened joints with silicone caulk.

Painting & Staining Siding

Exterior painting and staining projects are inescapable for do-it-yourselfers. If you've just completed a wood siding project, the job isn't finished until you seal the wood with a protective coat of stain or paint. Prep is important. Unless you prepare the surface by scraping, cleaning, and priming, you'll compromise the final coat and shorten the life of your paint or stain—in some cases dramatically.

Maybe you're simply intending to repaint or add a fresh coat of stain, but the current finish shows signs of problems. What do you do to correct the situation? This section will help you identify paint and stain problems, prepare surfaces properly, and give you application options to make the job easier and faster. Painting and staining are relatively inexpensive ways to keep your home's exterior looking its best.

High-quality painting tools, primers, and paints usually produce better results with less work than less-expensive products. The return on an investment in quality products is a project that goes smoothly and results in an attractive, durable paint job.

Plan to prime all unpainted surfaces and any surfaces or patches that have been stripped or worn down to bare wood. For bare wood, the best approach is to apply one primer coat followed by two top coats. But if the surface was previously painted and the old paint is still good, one coat of new paint is enough.

Although removing layers of old paint can be quite a chore, the proper materials can make the task go faster. If you don't own all of the specialty tools you need, such as a siding sander and heat gun, you can rent them at some home improvement stores and most rental centers. As you plan your painting project, make a list of the tools and materials you'll need.

Buy or rent a pressure washer and attachments to clean siding thoroughly and remove loose, flaky paint. Make sure to get the right unit. One with less than 1,200 psi won't do a good job, and one with more than 2,500 psi could damage your siding.

Tools for applying paint include roller and sleeve with ⅜" nap (A), corner roller (B), roller with ⅝" nap (C), 4" paintbrush (D), 3" paintbrush (E), 2" sash brush (F), 3"-wide roller (G). *Note: All brushes shown have synthetic bristles for use with latex-based paint.*

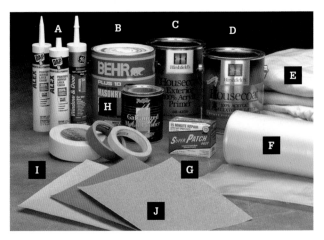

Materials for painting include painter's caulk (A), masonry, stucco, and brick paint (B), primer (C), house paint (D), drop cloth (E), plastic sheeting (F), epoxy wood filler (G), metal primer (H), masking tape (I), and 80-, 120-, and 150-grit sandpaper (J).

Estimating Paint ▸

To estimate the amount of paint you need for one coat:

- Calculate the square footage of the walls (length × height), the square footage of the soffit panels and trim that will be painted, and add 15% waste allowance.
- Subtract from this figure the square footage of doors and windows.
- Check the paint coverage rate listed on the label (350 sq. ft. per gallon is average).
- Divide the total square footage by the paint coverage rate to determine the number of gallons you need.

Tools for paint removal include ⅓ sheet finish sander (A), drill with wire-wheel attachments (B), ¼ sheet palm sander (C), heat gun (D), caulk gun (E), steel wool (F), wire brushes (G), stiff-bristle brush (H), sanding blocks (I), paint scraper (J), paint zipper (K), painter's 5-in-1 tool (L), detail scraper (M), and putty knife (N).

Paint-spraying equipment includes spray gun (A), hose (B), and compressor (C). Proper preparation requires plastic sheeting (D), masking tape (E), and drop cloths (F). Always use the necessary protective devices, including dual-cartridge respirator (G), and safety goggles (H).

Identifying Exterior Paint Problems

Two enemies work against painted surfaces—moisture and age. A simple leak or a failed vapor barrier inside the house can ruin even the finest paint job. If you notice signs of paint failure, such as blistering or peeling, take action to correct the problem right away. If the surface damage is discovered in time, you may be able to correct it with just a little bit of touch-up painting.

Evaluating the painted surfaces of your house can help you identify problems with siding, trim, roofs, and moisture barriers. The pictures on these two pages show the most common forms of paint failure and how to fix them. Be sure to fix any moisture problems before repainting.

Evaluate exterior painted surfaces every year, starting with areas sheltered from the sun. Paint failure will appear first in areas that receive little or no direct sunlight and is a warning sign that similar problems are developing in neighboring areas.

Common Forms of Paint Failure

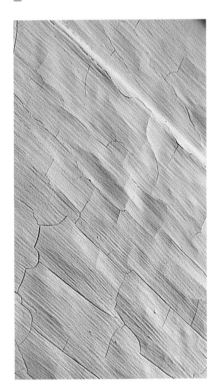

Blistering appears as a bubbled surface. It results from poor preparation or hurried application of primer or paint. The blisters indicate trapped moisture is trying to force its way through the surface. To fix isolated spots, scrape and touch up. For widespread damage, remove paint down to bare wood, then apply primer and paint.

Peeling occurs when paint falls away in large flakes. It's a sign of persistent moisture problems, generally from a leak or a failed vapor barrier. If the peeling is localized, scrape and sand the damaged areas, then touch up with primer and paint. If it's widespread, remove the old paint down to bare wood, then apply primer and paint.

Alligatoring is widespread flaking and cracking, typically seen on surfaces that have many built-up paint layers. It can also be caused by inadequate surface preparation or by allowing too little drying time between coats of primer and paint. Remove the old paint, then prime and repaint.

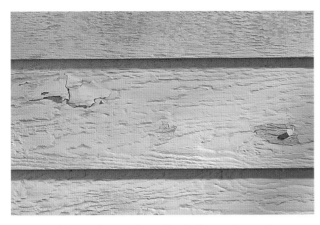

Localized blistering and peeling indicates that moisture, usually from a leaky roof, gutter system, or interior pipe, is trapped under the paint. Find and eliminate the leak, then scrape, prime, and repaint the area.

Clearly defined blistering and peeling occurs when a humid room has an insufficient vapor barrier. If there's a clear line where an interior wall ends, remove the interior wall and replace the vapor barrier.

Mildew forms in cracks and in humid areas that receive little direct sunlight. Wash mildewed areas with a 1:1 solution of household chlorine bleach and water or with trisodium phosphate (TSP).

Rust occurs when moisture penetrates paint on iron or steel. Remove the rust and loose paint with a drill and wire brush attachment, then prime and repaint.

Bleeding spots occur when nails in siding begin to rust. Remove the nails, sand out the rust, then drive in galvanized ring-shank nails. Apply metal primer, then paint to blend in with the siding.

Efflorescence occurs in masonry when minerals leech through the surface, forming a crystalline or powdery layer. Use a scrub brush and a muriatic acid solution to remove efflorescence before priming and painting.

Preparing to Paint

The key to an even paint job is to work on a smooth, clean, dry surface—so preparing the surface is essential. Generally, the more preparation work you do, the smoother the final finish will be and the longer it will last.

For the smoothest finish, sand all the way down to the bare wood with a power sander. For a less time-consuming (but rougher) finish, scrape off any loose paint, then spot-sand rough areas. You can use pressure washing to remove some of the flaking paint, but by itself, pressure washing won't create a smooth surface for painting.

Tools & Materials ▸

Pressure washer
Scraper
Sander
Sanding block
Putty knife
Stiff-bristle brush
Wire brush
Steel wool
Coarse abrasive pad
Drill
Wire-wheel attachment
Caulk gun
Heat gun
Proper respiratory
  protection
Sandpaper (80-,
  120-, 150-grit)
Putty
Paintable
  siliconized caulk
Muriatic acid
Sealant

Sanded to bare wood

Scraped and spot-sanded

Pressure washed only

The amount of surface preparation you do will largely determine the final appearance of your paint job. Decide how much sanding and scraping you're willing to do to obtain a finish you'll be happy with.

How to Remove Paint

1

Use a heat gun to loosen thick layers of old paint. Aim the gun at the surface, warm the paint until it starts to bubble, then scrape the paint as soon as it releases.

2

To remove large areas of paint on wood lap siding, use a siding sander with a disk that's as wide as the reveal on your siding.

How to Prepare Surfaces for Paint

Clean the surface and remove loose paint by pressure washing the house. As you work, direct the water stream downward, and don't get too close to the surface with the sprayer head. Allow all surfaces to dry thoroughly before continuing.

Scrape off loose paint using a paint scraper. Be careful not to damage the surface by scraping too hard.

Smooth out rough paint with a finishing sander and 80-grit sandpaper. Use sanding blocks and 80- to 120-grit sandpaper to sand hard-to-reach areas of trim. *Tip: You can make sanding blocks from dowels, wood scraps, or garden hoses.*

Use detail scrapers to remove loose paint in hard-to-reach areas. Some of these scrapers have interchangeable heads that match common trim profiles.

Inspect all surfaces for cracks, rot, and other damage. Mark affected areas with colored pushpins or tape. Fill the holes and cracks with epoxy wood filler.

Use a finishing sander with 120-grit sandpaper to sand down repaired areas, ridges, and hard edges left from the scraping process, creating a smooth surface.

How to Prepare Window & Door Trim for Paint

Scuff-sand glossy surfaces on doors, window casings, and all surfaces painted with enamel paint. Use a coarse abrasive pad or 150-grit sandpaper.

Fill cracks in siding and gaps around window and door trim with paintable siliconized acrylic caulk.

How to Remove Clear Finishes

Pressure wash stained or unpainted surfaces that have been treated with a wood preservative or protectant before recoating them with fresh sealant.

Use a stiff-bristle brush to dislodge any flakes of loosened surface coating that weren't removed by pressure washing. Don't use a wire brush on wood surfaces.

How to Prepare Metal & Masonry for Paint

1

Remove rust and loose paint from metal hardware, such as railings and ornate trim, using a wire brush. Cover the surface with metal primer immediately after brushing to prevent the formation of new rust.

2

Scuff-sand metal siding and trim with medium-coarse steel wool or a coarse abrasive pad. Wash the surface and let dry before priming and painting.

3

Remove loose mortar, mineral deposits, or paint from mortar lines in masonry surfaces with a drill and wire-wheel attachment. Clean broad, flat masonry surfaces with a wire brush. Correct any minor damage before repainting.

4

Dissolve rust on metal hardware with diluted muriatic acid solution. When working with muriatic acid, it's important to wear safety equipment, work in a well-ventilated area, and follow all manufacturer's directions and precautions.

Applying Paint & Primer

Schedule priming and painting tasks so that you can paint within two weeks of priming surfaces. If more than two weeks pass, wash the surface with soap and water before applying the next coat.

Check the weather forecast and keep an eye on the sky while you work. Damp weather or rain within two hours of application will ruin a paint job. Don't paint when the temperature is below 50° or above 90°F. Avoid painting on windy days—it's dangerous to be on a ladder in high winds, and wind blows dirt onto the fresh paint.

Plan each day's work so you can follow the shade. Prepare, prime, and paint one face of the house at a time, and follow a logical painting order. Work from the top of the house down to the foundation, covering an entire section before you move the ladder or scaffolding.

Tools & Materials ▸

Paintbrush (4", 2½", 3")   Primer
Sash brush   House paint
Scaffolding   Trim paint
Ladders   Cleanup materials

Paint in a logical order, starting from the top and working your way down. Cover as much surface as you can reach comfortably without moving your ladder or scaffolding. After the paint or primer dries, touch up any unpainted areas that were covered by the ladder or ladder stabilizer.

Tips for Applying Primer & Paint

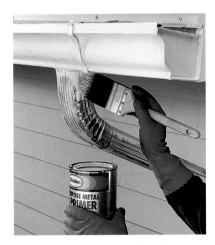

Use the right primer and paint for each job. Always read the manufacturer's recommendations.

Plan your painting sequence so you paint the walls, doors, and trim before painting stairs and porch floors. This prevents the need to touch up spills.

Apply primer and paint in the shade or indirect sunlight. Direct sun can dry primers and paints too quickly and trap moisture below the surface, which leads to blistering and peeling.

Tips for Selecting Brushes & Rollers

Wall brushes, which are thick, square brushes 3 to 5" wide, are designed to carry a lot of paint and distribute it widely. *Tip: It's good to keep a variety of clean brushes on hand, including 2½", 3", and 4" flat brushes, 2" and 3" trim brushes, and tapered sash brushes.*

Trim and tapered sash brushes, which are 2 to 3" wide, are good for painting doors and trim, and for cutting in small areas.

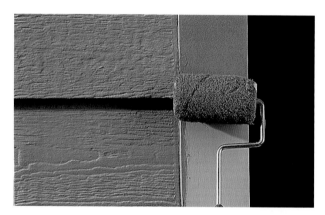

Paint rollers work best for quickly painting smooth surfaces. Use a roller sleeve for broad surfaces.

Use a 3" roller to paint flat-surfaced trim, such as end caps and corner trim.

Tips for Loading & Distributing Paint

Load your brush with the right amount of paint for the area you're covering. Use a full load of paint for broad areas, a moderate load for smaller areas and feathering strokes, and a light load when painting or working around trim.

Hold the brush at a 45° angle and apply just enough downward pressure to flex the bristles and squeeze the paint from the brush.

How to Use a Paintbrush

Load the brush with a full load of paint. Starting at one end of the surface, make a long, smooth stroke until the paint begins to feather out. *Tip: Paint color can vary from one can to the next. To avoid problems, pour all of your paint into one large container and mix it thoroughly. Pour the mixed paint back into the individual cans and seal them carefully. Stir each can before use.*

At the end of the stroke, lift the brush without leaving a definite ending point. If the paint appears uneven or contains heavy brush marks, smooth it out without overbrushing.

Reload the brush and make a stroke from the opposite direction, painting over the feathered end of the first stroke to create a smooth, even surface. If the junction of the two strokes is visible, rebrush with a light coat of paint. Feather out the starting point of the second stroke.

Tips for Using Paint Rollers

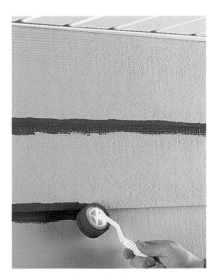

Wet the roller nap, then squeeze out the excess water. Position a roller screen inside a 5-gal. bucket. Dip the roller into the paint, then roll it back and forth across the roller screen. The roller sleeve should be full but not dripping when lifted from the bucket.

Cone-shaped rollers work well for painting the joints between intersecting surfaces.

Doughnut-shaped rollers work well for painting the edges of lap siding and moldings.

How to Paint Fascia, Soffits & Trim

Prime all surfaces to be painted, and allow ample drying time. Paint the face of the fascia first, then cut in paint at the bottom edges of the soffit panels. *Tip: Fascia and soffits are usually painted the same color as the trim.*

Paint the soffit panels and trim with a wide brush. Start by cutting in around the edges of the panels using the narrow edge of the brush, then feather in the broad surfaces of the soffit panels with full loads of paint. Be sure to get good coverage in the grooves.

Paint any decorative trim near the top of the house at the same time you paint the soffits and fascia. Use a 2½" or 3" paintbrush for broader surfaces, and a sash brush for more intricate trim areas.

How to Paint Siding

Paint the bottom edges of lap siding by holding the paintbrush flat against the wall. Paint the bottom edges of several siding pieces before returning to paint the faces of the same boards.

Paint the broad faces of the siding boards with a 4" brush using the painting technique shown on page 210. Working down from the top of the house, paint as much surface as you can reach without leaning beyond the sides of the ladder.

Paint the siding all the way down to the foundation, working from top to bottom. Shift the ladder or scaffolding, then paint the next section. *Tip: Paint up to the edges of end caps and window or door trim that will be painted later.*

On board and batten or vertical panel siding, paint the edges of the battens, or top boards, first. Paint the faces of the battens before the sides dry, then use a roller with a ⅝"-nap sleeve to paint the large, broad surfaces between the battens.

How to Paint Stucco Walls

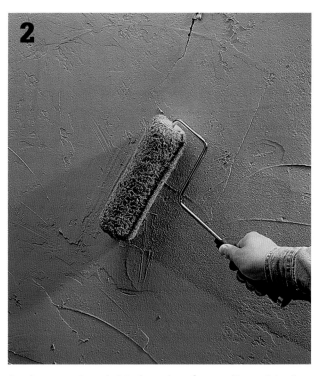

Using a large paintbrush, paint the foundation with antichalking masonry primer and let it dry. Using concrete paint and a 4" brush, cut in the areas around basement windows and doors.

Apply concrete paint to board surfaces with a paint roller and a ⅝"-nap sleeve. Use a 3" trim roller or a 3" paintbrush for trim.

Tips for Cleaning Painting Tools

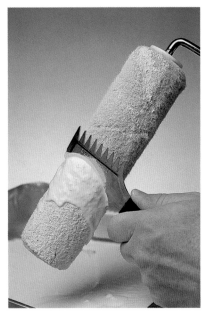

Scrape paint from roller covers with the curved side of a cleaner tool.

Use a spinner tool to remove paint and solvent from brushes and roller covers.

Comb brushes with the spiked side of a cleaner tool to properly align bristles for drying.

Cleaning Wood Siding ▸

Wood siding can provide a long and graceful service life, but it does require some routine maintenance to keep it looking its best. If your siding has a tinted stain finish, the effects of sunlight will slowly cause the color to fade. Eventually, the stain will lose its protective qualities, and the wood will turn silvery gray. Areas of your siding that do not receive direct sunlight are still prone to other problems. Shady areas can remain damp, which invites mold or algae growth and eventually rot.

The best way to combat fading and deterioration is to thoroughly clean your wood siding every couple of years and apply a fresh coat of stain. If your siding is faded but not discolored, you may be able to simply wash it down with soapy water to remove dirt and other debris, then stain it again. If the wood has turned gray but isn't deteriorated, use a commercial deck cleaner, followed by thorough washing to restore the original wood color. If there are signs of mold or algae growth, use a diluted mixture of trisodium phosphate or a biodegradable cleaning product applied with a pump sprayer to kill the growth. Scrub these areas thoroughly, then wash off the chemicals.

Before you begin, cover shrubbery, electrical outlets, and leaky doors or windows with plastic sheeting to protect them. Dampen plants under the sheeting to keep them from overheating in the sun. When using chemicals or pressure washing, be sure to wear safety glasses and protective clothing if required. Pressure wash judiciously using the proper wand tip. A pressure washer is an excellent cleaning tool, but it has the power to damage your siding if you're not careful.

Tools & Materials ▸

Garden hose and spray applicator or nozzle	Stiff-bristle push broom	Plastic sheeting	Trisodium phosphate granules
Hand pump sprayer	Pressure washer (optional)	Deck cleaner	or biodegradable cleaner

Cleaning Tips

For routine cleaning: If your stain color has faded but the siding is in good condition, use a garden hose to wash off dirt or other debris. You can use a spray applicator and diluted mixture of mild detergent to clean off minor staining. Allow the siding to dry thoroughly for a day or two before restaining.

To revitalize weathered wood: You can usually restore the natural wood color of sun-faded cedar or redwood siding using a commercial deck cleaner formulated with a brightening agent. Apply it with a hand pump sprayer and let it soak in according to the manufacturer's instructions. You may need to follow the application by scrubbing the siding with a stiff-bristle push broom to work the chemicals into the wood.

Using Paint-Spraying Equipment

Spray equipment can make quick work of painting, but it still requires the same careful preparation work as traditional brush and roller methods. Part of that prep work involves using plastic to completely cover doors, windows, and other areas that you don't want painted, rather than just taping them off.

Spray equipment can be purchased or rented at hardware and home improvement stores. There are several types and sizes of spray equipment, including high-volume low-pressure (HVLP), airless, air-assisted airless, and electrostatic enhanced. They all work the same way—by atomizing paint and directing it to a worksurface in a spray or fan pattern. For our project, we used an HVLP sprayer, which we recommend because it produces less overspray and more efficient paint application than other sprayers.

Be sure to read and follow all safety precautions for the spray equipment. Since the paint is under a lot of pressure, it can not only tear the skin, but it can inject toxins into the bloodstream if used incorrectly. Wear the proper safety protection, such as safety glasses and a respirator, when spray painting the house.

As with other paint applications, pay close attention to the weather. Don't spray if rain is likely, and don't spray on windy days, since the wind can carry the paint particles away from the siding.

Tools & Materials ▸

Utility knife	Masking tape
Spray equipment	Plastic
Paint	Cardboard
Safety glasses	Cheesecloth
Respirator	5-gal. bucket

Paint sprayers allow you to cover large areas of siding and trim in a short amount of time. They also make it easier to paint areas that are hard to reach with a brush or roller.

How to Paint Using a Paint Sprayer

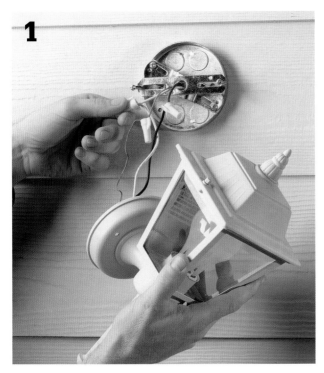

Remove outside light fixtures, window and door screens, and other detachable items that you don't want painted. Turn off the circuit breaker before disconnecting power to a fixture.

Cover doors, windows, and any other areas you don't want painted using plastic and masking tape.

Strain the paint through cheesecloth to remove particles and debris. Mix the paint together in a 5-gal. bucket. Fill the sprayer container.

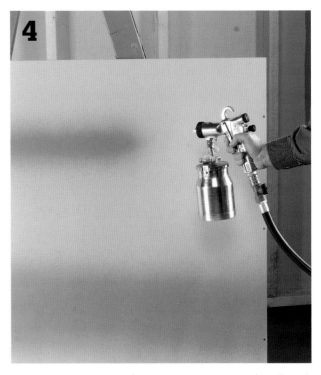

Spray a test pattern of paint on a scrap piece of cardboard. Adjust the pressure until you reach an even "fan" without any thick lines along the edge of the spray pattern.

(continued)

5

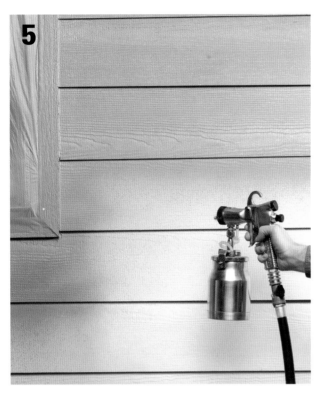

Cut in around doors and windows with the paint. Spray the paint along each side of the doors and windows, applying the paint evenly.

6

If you happen to spray an excessive amount of paint in an area and it starts to run, stop the sprayer. Use a paintbrush to spread out the paint and eliminate the runs.

7

Hold the spray gun perpendicular to the house, approximately 12" from the wall. Start painting near the top of the wall, close to a corner. Move your entire arm, rather than just the wrist, in a steady side-to-side motion. Do not wave your arm in an arc. Start your arm movement, then start the gun.

8

Spray the paint in an even motion, being careful not to tilt the gun. As you sweep your arm back and forth, overlap each coat of paint by 20 to 30 percent, working your way down the wall. When stopping, release the trigger before discontinuing your motion.

How to Paint Doors Using a Paint Sprayer

1

Remove the door by taking off the hinges. Remove all hardware from the door, such as handles and locks. If the door contains glass, you can either tape it off, or allow paint to get on the glass and then scrape it off with a razor after it's dry.

2

Prop up the door so it stands vertically. Starting at the top of the door, spray on the paint. As you make passes across the door, slightly go past the edges before sweeping back in the opposite direction. Wait until the paint is completely dry, then turn the door around and paint the other side.

Evaluating Siding & Trim

The first step in inspecting and evaluating siding and trim is to identify the type of material used on the house. Once you determine the material, take a close look at the problem area and determine the best method to fix it. If your siding is still under warranty, read through the warranty document before starting any repairs. Making repairs yourself could invalidate the product warranty. If the siding was professionally installed, you may want to talk to your contractor about the repairs.

In addition to looking unsightly, small siding problems can escalate into larger and more costly problems. As soon as you spot any siding damage, take steps to fix it immediately, especially if there's a possibility of water infiltration.

Evaluate broad trim pieces, such as the end cap trim shown above, and make repairs using the same techniques as for siding.

Blistering or peeling paint can be scraped off, then the wood can be painted.

Damaged panels of siding can be removed and replaced; you don't need to replace the entire siding. The new panels are then painted to match the surrounding ones.

Common Siding Problems

Separated joints can occur in any type of lap siding, but they're most common in wood lap. Gaps between ⅛ and ¼" thick can be filled with caulk. Gaps ⅜" or wider could mean that your house has a serious moisture or shifting problem. Consult a building inspector.

Buckling occurs most frequently in manufactured siding when expansion gaps are too small at the points where the siding fits into trim and channels. If possible, move the channels slightly to give the siding more room. If not, remove the siding, trim the length slightly, then reinstall.

Minor surface damage to metal siding is best left alone in most cases—unless the damage has penetrated the surface. With metal products, cosmetic surface repairs often look worse than the damage.

Missing siding, such as cedar shakes that have been blown away from the wall, should be replaced immediately. Check the surrounding siding to make sure it's secure.

Repairing Siding

Damage to siding is fairly common, but fortunately, it's also easy to fix. Small to medium holes, cracks, and rotted areas can be repaired with filler or by replacing the damaged sections with matching siding.

If you cannot find matching siding for repairs at building centers, check with salvage yards or siding contractors. When repairing aluminum or vinyl siding, contact the manufacturer or the contractor who installed the siding to help you locate matching materials and parts. If you're unable to find an exact match, remove a section of original siding from a less visible area of the house, such as the back of the garage, and use it for the patch. Cover the gap in the less visible area with a close matching siding, where the mismatch will be less noticeable.

As a last resort, you can paint aluminum or vinyl siding with high quality acrylic primer and paint. Bring a piece of original siding to the paint store for an exact color match.

Tools & Materials ▶

Aviation snips	Paintbrush
Caulk gun	Epoxy wood filler
Drill	Epoxy glue
Flat pry bar	Galvanized ring-
Hammer	shank siding nails
Straightedge	Siliconized
Tape measure	acrylic caulk
Utility knife	Roofing cement
Zip-lock tool	30# felt paper
Chisel	Sheathing
Trowel	Trim
Screwdrivers	Replacement siding
Hacksaw	End caps
Circular saw	Wood preservative
Jigsaw	Primer
Key hole saw	Paint or stain
Nail set	Metal sandpaper
Stud finder	

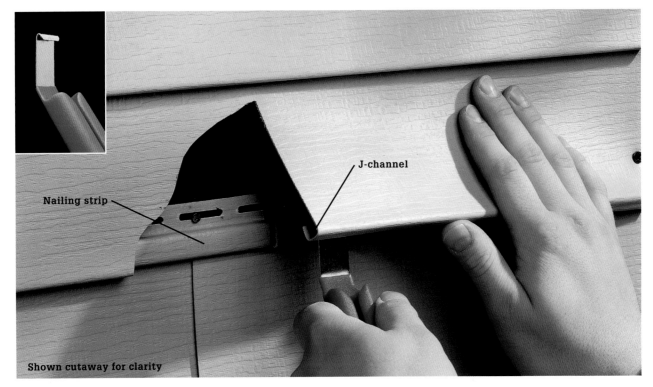

Nailing strip

J-channel

Shown cutaway for clarity

Vinyl and metal siding panels have a locking J-channel that fits over the bottom of the nailing strip on the underlying piece. Use a zip-lock tool (inset) to separate panels. Insert the tool at the seam nearest the repair area. Slide it over the J-channel, pulling outward slightly, to unlock the joint from the siding below.

How to Patch Vinyl Siding

Starting at the seam nearest the damaged area, unlock interlocking joints using a zip-lock tool. Insert spacers between the panels, then remove the fasteners in the damaged siding using a flat pry bar. Cut out the damaged area using aviation snips. Cut a replacement piece 4" longer than the open area, and trim 2" off the nailing strip from each end. Slide the piece into position.

Insert siding nails in the nailing strip, then position the end of a flat pry bar over each nail head. Drive the nails by tapping on the neck of the pry bar with a hammer. Place a scrap piece of wood between the pry bar and siding to avoid damaging the siding. Slip the locking channel on the overlapping piece over the nailing strip of the replacement piece. *Tip: If the damaged panel is near a corner, door, or window, replace the entire panel. This eliminates an extra seam.*

How to Patch Aluminum Siding

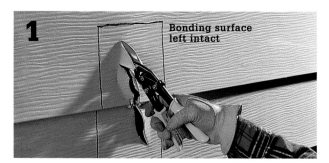

Cut out the damaged area or leave it if it's flat. Use a razor knife or aviation snips. Leave an exposed area on top of the uppermost piece to act as a bonding surface. Cut a patch 4" larger than the repair area. Remove the nailing strip. Smooth the edges with metal sandpaper.

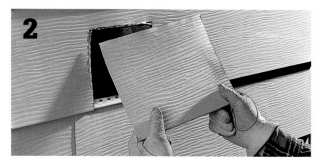

Nail the lower patch in place by driving siding nails through the nailing flange. Apply clear silicone caulk to the back of the top piece, then press it into place, slipping the locking channel over the nailing strip of the underlying piece. Caulk the seams.

How to Replace Aluminum End Caps

Remove the damaged end cap. If necessary, pry the bottom loose, then cut along the top with a hacksaw blade. Starting at the bottom, attach the replacement end caps by driving siding nails through the nailing tabs and into the framing members.

Trim the nailing tabs off the top replacement cap. Apply roofing cement to its back. Slide the cap over the locking channels of the siding panels. Press the top cap securely in place.

How to Replace Board & Batten Siding

Remove the battens over the damaged boards. Pry out the damaged boards in their entirety. Inspect the underlying housewrap, and patch if necessary.

Cut replacement boards from the same type of lumber, allowing a ⅛" gap at the side seams. Prime or seal the edges and the back side of the replacement boards. Let them dry.

Nail the new boards in place using ring-shank siding nails. Replace the battens and any other trim. Prime and paint or stain the new boards to blend with the surrounding siding.

How to Replace Wood Shakes & Shingles

Split damaged shakes or shingles with a hammer and chisel, and remove them. Insert wood spacers under the shakes or shingles above the repair area, then slip a hacksaw blade under the top board to cut off any remaining nail heads.

Cut replacement shakes or shingles to fit, leaving a ⅛- to ¼"-wide gap at each side. Coat all sides and edges with wood preservative. Slip the patch pieces under the siding above the repair area. Drive siding nails near the top of the exposed area on the patches. Cover nail heads with caulk. Remove the spacers.

How to Replace Lap Siding

If the damage is caused by water, locate and repair the leak or other source of the water damage.

Mark the area of siding that needs to be replaced. Make the cutout lines over the center of the framing members on each side of the repair area, staggering the cuts to offset the joints. *Tip: Use an electronic stud finder to locate framing members, or look for the nail heads.*

Insert spacers beneath the board above the repair area. Make entry cuts at the top of the cutting lines with a key hole saw, then saw through the boards and remove them. Pry out any nails or cut off the nail heads using a hacksaw blade. Patch or replace the sheathing and building paper, if necessary.

Measure and cut replacement boards to fit. Use the old boards as templates to trace cutouts for fixtures and openings. Use a jigsaw to make the cutouts. Apply wood sealer or primer to the ends and backs of the boards. Let them dry.

Nail the new boards in place with siding nails, starting with the lowest board in the repair area. At each framing member, drive nails through the bottom of the new board and the top of the board below. *Tip: If you removed the bottom row of siding, nail a 1 × 2 starter strip along the bottom of the patch area.*

Fill joints with caulk (use paintable caulk for painted wood or tinted caulk for stained wood). Prime and paint or stain the replacement boards to match the surrounding siding.

Repairing Stucco Walls

Although stucco siding is very durable, it can be damaged, and over time it can crumble or crack. The directions given below work well for patching small areas less than 2 sq. ft. For more extensive damage, the repair is done in layers, as shown on the opposite page.

Tools & Materials ▶

Caulk gun	Metal primer
Disposable paintbrush	Stucco patching compound
Putty knife	Bonding adhesive
Mason's trowel	Denatured alcohol
Square-end trowel	Metal primer
Hammer	Stucco mix
Whisk broom	Masonry paint
Wire brush	1½" roofing nails
Masonry chisel	15# building paper
Aviation snips	Self-furring metal lath
Pry bar	Masonry caulk
Drill with masonry bit	Tint
Scratching tool	Metal stop bead

Fill thin cracks in stucco walls with masonry caulk. Overfill the crack with caulk and feather until it's flush with the stucco. Allow the caulk to set, then paint it to match the stucco. Masonry caulk stays semiflexible, preventing further cracking.

How to Patch Small Areas

Remove loose material from the repair area using a wire brush. Use the brush to clean away rust from any exposed metal lath, then apply a coat of metal primer to the lath.

Apply premixed stucco repair compound to the repair area, slightly overfilling the hole using a putty knife or trowel. Read manufacturer's directions, as drying times vary.

Smooth the repair with a putty knife or trowel, feathering the edges to blend into the surrounding surface. Use a whisk broom or trowel to duplicate the original texture. Let the patch dry for several days, then touch it up with masonry paint.

How to Repair Large Areas

Make a starter hole with a drill and masonry bit, then use a masonry chisel and hammer to chip away stucco in the repair area. *Note: Wear safety glasses and a particle mask or respirator when cutting stucco. Cut self-furring metal lath to size and attach it to the sheathing using roofing nails. Overlap pieces by 2". If the patch extends to the base of the wall, attach a metal stop bead at the bottom.*

To mix your own stucco, combine three parts sand, two parts Portland cement, and one part masonry cement. Add just enough water so the mixture holds its shape when squeezed (inset). Mix only as much as you can use in 1 hour. *Tip: Premixed stucco works well for small jobs, but for large ones, it's more economical to mix your own.*

Apply a ⅜"-thick layer of stucco directly to the metal lath. Push the stucco into the mesh until it fills the gap between the mesh and the sheathing. Score horizontal grooves into the wet surface using a scratching tool. Let the stucco dry for two days, misting it with water every 2 to 4 hours.

Apply a second, smooth layer of stucco. Build up the stucco to within ¼" of the original surface. Let the patch dry for two days, misting every 2 to 4 hours.

Combine finish coat stucco mix with just enough water for the mixture to hold its shape. Dampen the patch area, then apply the finish coat to match the original surface. Dampen the patch periodically for a week. Let it dry for several more days before painting.

SOFFITS, FASCIA, GUTTERS & VENTS

Soffits & Vents

An effective ventilation system equalizes temperatures on both sides of the roof, which helps keep your house cooler in the summer and prevents ice dams along the roof eaves in cold climates.

One strategy for increasing roof ventilation is to add more of the existing types of vents. Or, if you're reroofing, consider replacing all of your roof vents with a continuous ridge vent (pages 60 to 61). You can increase intake ventilation by adding more soffit vents. If you're replacing your soffits with aluminum soffits, install vented soffit panels that allow air intake (pages 230 to 233).

Determining Ventilation Requirements

Measure attic floor space to determine how much ventilation you need. You should have 1 sq. ft. each of intake and outtake ventilation for every 300 sq. ft. of unheated attic floor space.

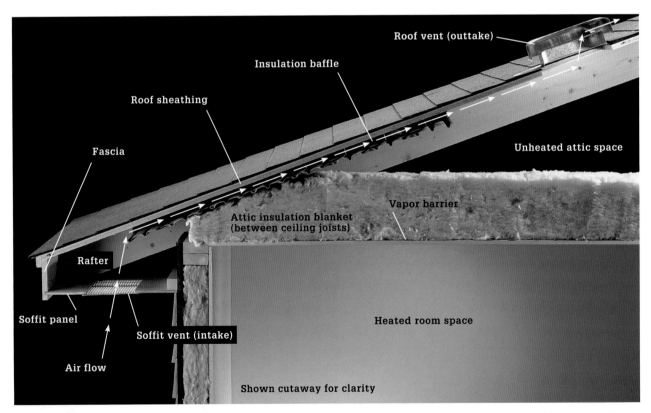

Roof vent (outtake)

Insulation baffle

Roof sheathing

Fascia

Unheated attic space

Vapor barrier

Attic insulation blanket (between ceiling joists)

Rafter

Soffit panel

Heated room space

Soffit vent (intake)

Air flow

Shown cutaway for clarity

Sufficient airflow prevents heat buildup in your attic, and it helps protect your roof from damage caused by condensation or ice. A typical ventilation system has vents in the soffits to admit fresh air, which flows upward beneath the roof sheathing and exits through the roof vents.

Types of Vents

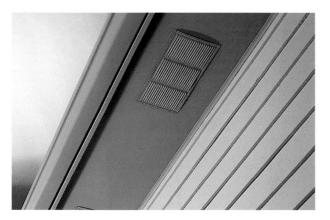

Soffit vents can be added to increase airflow into attics on houses with a closed soffit system.

Continuous soffit vents provide even airflow into attics. They are usually installed during new construction, but they can be added as retrofits to unvented soffit panels.

Roof vents can be added near the ridge line when you need to increase outtake ventilation. Fixed roof vents are easy to install and have no mechanical parts that can break down.

Vented soffit panels are used with aluminum soffits to allow airflow along the eaves.

Gable and dormer vents generally are installed to increase ventilation. The vents come in a variety of styles and colors to match the siding.

Continuous ridge vents create an even outtake airflow because they span the entire ridge. Barely noticeable from the ground, ridge vents can be added at any time.

Aluminum Soffits

Older soffits may be weathered or rotted and may not allow adequate airflow. If more than 15 percent of your soffits need to be repaired, your best option is to replace them. This project shows how to completely remove the old soffits and fascia and install aluminum soffits, which are maintenance free. If your old subfascia is in good condition, it will not need to be replaced.

The project starting on the opposite page details the installation of soffits on an eaves system that has rafter lookouts. The soffits are installed directly beneath these lookouts. If your eaves do not have rafter lookouts, follow the instructions starting on page 232. This project also shows how to install soffits around corners.

For both eaves systems, an F-channel serves as a mounting channel to hold the soffits in place along the house. You can also install the channel along the subfascia, as shown in step 4 on page 231, or you can nail the soffits directly to the subfascia, as shown in step 4 on page 233. Drive nail heads flush with the surface. Driving the nails too deep can knock the soffits out of shape and prevent movement. Since the soffits will receive addition nailing when the fascia is installed, you don't need to drive a nail in every V-groove in the soffits.

To cut soffits, use a circular saw with a fine-tooth blade installed backward. Don't cut all of your panels at the start of the job since the width will probably change slightly as you move across the house.

Use vented soffit panels to work in conjunction with roof or attic vents. This improves airflow underneath the roof, which prevents moisture damage and ice dams. Provide 1 sq. ft. of soffit vents for every 150 sq. ft. of unheated attic space. For a consistent appearance, make sure all of the fins on the soffit vents are pointed in the same direction.

Tools & Materials ▸

Flat pry bar	F-channel
Hammer	(mounting channel)
Circular saw with	1¼" aluminum
fine-tooth metal	trim nails
blade (installed	16d common nails
backward)	Nailing strips
Drill	Drip edge
Tape measure	2¼" deck screws
Aviation snips	8d box nails
Level	Subfascia, if needed
Framing square	(2 × 4, 1 × 8,
Soffit panels	or 2 × 8)
T-channel	

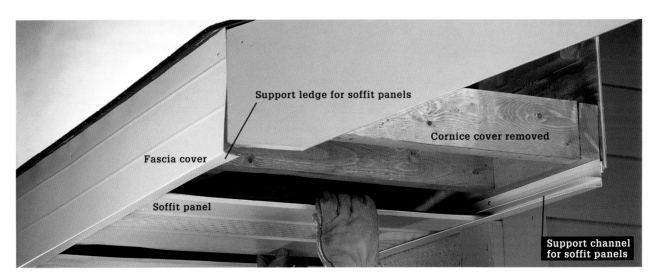

Labels: Support ledge for soffit panels • Cornice cover removed • Fascia cover • Soffit panel • Support channel for soffit panels

Install a new soffit system if your old system has failed, or if pests have infested the open eaves areas of your roof overhang. A complete soffit system consists of fabricated fascia covers, soffit panels (nonventilated or ventilated), and support channels that hold the panels at the sides of your house. Most soffit systems sold at building centers are made of aluminum.

How to Install Aluminum Soffits (with Rafter Lookouts)

Remove trim, soffits, and fascia along the eaves using a flat pry bar. If the eaves contain debris, such as bird nests or rotted wood, clean them out.

Check the rafters and rafter lookouts for decay or damage. Repair or replace them as needed.

Install new 1 × 8 or 2 × 8 subfascia over the rafters and rafter lookouts using 16d nails. Butt subfascia boards together at rafter or rafter lookout locations. Install drip edge at the top of the subfascia. Leave a ¹⁄₁₆" gap between the drip edge and the subfascia for the fascia to fit.

Install F-channels for the soffit panels along the bottom inside edge of the subfascia and along the outside wall of the house directly below the rafter lookouts. If more than one piece of channel is needed, butt pieces together.

(continued)

If the soffit panels will span more than 16", or if your house is subjected to high winds, add nailing strips to provide additional support.

Measure the distance between the mounting channels, subtract ⅛", and cut soffits to size. Slide the soffit panels in place, fitting the ends inside the mounting channels. Nail the panels to the nailing strips, if you've installed them.

Install soffit panels in the remaining spaces, cutting them to fit as needed. When finished, install the fascia (pages 234 to 235).

How to Install Aluminum Soffits (without Rafter Lookouts)

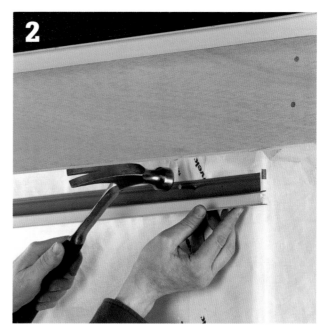

Remove the old soffits and fascia, following step 1 on page 231. Place a level at the bottom of the subfascia board, level across to the house, and make a mark. Measure down from the mark a distance equal to the thickness of the soffits (usually about ¼"). Do this on each end of the wall. Snap a chalkline between the lower marks.

Start the F-channel at a corner and align the bottom edge with the chalkline. Nail the channel to the wall at stud locations using 8d box nails. If more than one F-channel is needed, butt the pieces together.

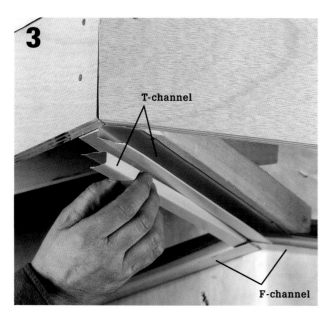

At corners, cut a 2 × 4 to fit between the house and the inside corner of the subfascia to provide support for the T-channel. Notch the 2 × 4 as needed, then nail in place so when the T-channel is installed, it will be aligned with the F-channel. Cut the T-channel to fit. Place it against the 2 × 4, setting the back edge inside the F-channel, and nail in place.

Measure between the F-channel and the outside edge of the subfascia. Subtract ¼" and cut the soffits to size. For corners, miter the panels to fit the T-channel. Install the first panel inside the channel. Make sure the panel is square to the subfascia using a framing square. Nail the panel to the subfascia at the V-grooves. Slide the next panel against the first, locking them together. Nail the panel in place. Install remaining panels the same way.

Variations for Installing Aluminum Soffits

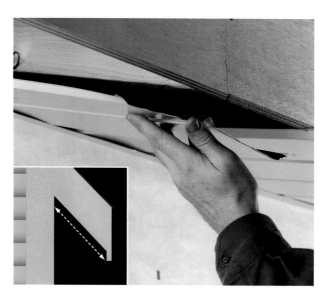

Straight corners are made by installing the T-channel parallel with one of the F-channels. Align the outside edge of the T-channel with the outside edge of the installed F-channel. Keep the T-channel back ¼" from the outside of the subfascia, and nail it in place. Install the soffits in the channels.

Inclined overhangs allow soffits to run the same angle as the rafters. At the end of the rafter overhangs, measure from the bottom of the rafter to the bottom of the subfascia. Add the thickness of the soffits, then measure down from the rafters along the wall and make a mark at this distance. Do this on each end of the wall. Snap a chalkline between the marks. Align the bottom of the F-channel with the chalkline, nail the channel to the wall, then install the soffits.

Aluminum Fascia

Fascia fits under the drip edge and against the subfascia to provide a smooth transition from the roof to the eaves. You may need to temporarily remove any nails in the face of the drip edge so the fascia can slide in behind it. If your roof does not have drip edge, install a finish trim, such as undersill, at the top of the subfascia to receive the fascia.

If you're also replacing your gutters, take down the gutters first, then install the fascia. If you don't want to remove the gutters, you can slip the fascia behind them while they're in place.

Fascia is nailed along the lip covering the soffits, and the top is held in place by the drip edge, so it doesn't require any face nailing.

Tools & Materials ▸

Hammer
Aviation snips
Tape measure
Chalkline
Fascia
Aluminum trim nails

The fascia is installed over the subfascia to cover the exposed edges of the soffits and enhance the appearance of your home. The fascia is usually the same color and material as your soffits.

How to Install Aluminum Fascia

Remove the old fascia, if necessary. Measure from the top of the drip edge to the bottom of the soffits, and subtract ¼". Cut the fascia to this measurement by snapping a chalkline across the face and cutting with aviation snips. (This cut edge will be covered by the drip edge.) *Tip: If your old fascia is wood and still in good shape, you can install aluminum fascia over it without removing it.*

Slide the cut edge of the fascia behind the drip edge. Place the bottom lip over the soffits. Make sure the fascia is tight against the soffits and against the subfascia, then nail through the lip into the subfascia. Nail approximately every 16" at a V-groove location in the soffits.

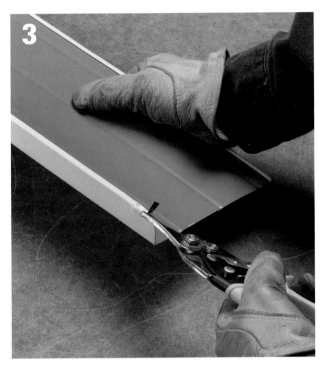

To overlap fascia panels, cut the ridge on the lip of the first panel 1" from the end using aviation snips. Place the second panel over the first, overlapping the seam by 1". Nail the fascia in place.

At outside corners, cut the lip and top edge of the first panel 1" from the end. Place a piece of wood 1" from the end, and bend the panel to form a 90° angle. Install the panel at the corner. Cut a 45° angle in the lip of the second panel. Align the end of this panel with the corner, overlapping the first panel.

For inside corners, cut and bend the first panel back 1" from the end to make a tab. Install the panel. On the second panel, cut a 45° angle in the lip. Slide the panel over the first panel, butting the end against the adjacent fascia. Nail the panel in place.

Install soffit panels to close off the area between the fascia cover and the exterior wall (see pages 230 to 231).

Wood Soffits

Wood soffits are typically used on houses with wood or fiber-cement siding, and are painted the same color as the trim. To see how the soffits fit in relation to the fascia and rafters, refer to the photo on page 238. You can use plywood or engineered wood for the soffits. Engineered wood has the advantage of being treated to resist termites and fungus, and it's more resistant to warping and shrinking. Plywood has the advantage of being less expensive. If your soffits are more than 24" wide, install a nailing strip between panels to hold the seams tightly together.

When replacing soffits, you may need to remove the top course of siding and any trim pieces under the old soffits before starting the new installation. Remove the pieces carefully, then reinstall them once the soffit job is finished.

Tools & Materials ▶

Hammer	2 × 2 lumber
Circular saw	Acrylic latex caulk
Level	Vents
Chalkline	Primer
Caulk gun	Paint
Paintbrush	
Drill	
Jigsaw	
⅜" plywood	
16d box nails	
6d corrosion-resistant	
nails	

Wood soffits cover the eaves area between the fascia and siding. Painted soffit moldings beneath the soffits give the house a finished look. Soffit vents installed at regular intervals play a vital role in the home's ventilation system.

How to Install Wood Soffits

1

Hold a level against the bottom edge of the subfascia and level across to make a mark on the wall. Do this on both ends of the wall, then snap a chalkline between the marks.

2

Align the bottom edge of 2 × 2 lumber with the chalkline. Nail the lumber to wall studs using 16d nails.

3

Measure the distance from the wall to the outside of the subfascia, subtract ¼", and rip the soffits to this width. Apply primer to the soffits. If using wood that's already primed, apply primer to the cut edges only.

4

Place the soffit against the 2 × 2 and subfascia, staying ⅛" from the edges. Nail in place using 6d nails. Install remaining soffits, keeping a ⅛" gap between panels. Caulk the gaps between soffits and between the soffits and wall. Paint soffits as desired. Let the paint dry.

5

Mark the vent locations in the soffits by holding the vent in place and tracing around it. Drill starter holes at opposite corners of the outline, then cut out the opening with a jigsaw. Install the vent using the fasteners that came with it. Do this at each vent location.

Variation: If the soffits have rafter lookouts, you don't need to install 2 × 2s. Instead, nail the soffits directly to the rafter headers and lookouts. Make sure soffit seams fall midway across rafter lookouts.

Repairing Wood Fascia & Soffits

Fascia and soffits add a finished look to your roof and promote a healthy roof system. A well-ventilated soffit system prevents moisture from building up under the roof and in the attic.

Most fascia and soffit problems can be corrected by cutting out sections of damaged material and replacing them. Joints between fascia boards are lock nailed at rafter locations, so you should remove whole sections of fascia to make accurate bevel cuts for patches. Soffits can often be left in place for repairs.

Tools & Materials ▸

Circular saw
Jigsaw
Drill
Putty knife
Hammer
Flat pry bar
Nail set
Chisel
Caulk gun
Paintbrush

Replacement materials
Nailing strips
Galvanized deck screws (2", 2½")
4d galvanized casing nails
Acrylic caulk
Primer
Paint

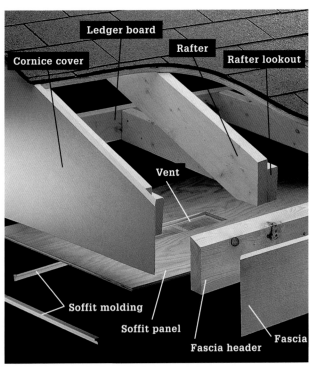

Fascia and soffits close off the eaves area beneath the roof overhang. The fascia covers the ends of rafters and rafter lookouts, and provides a surface for attaching gutters. Soffits are protective panels that span the area between the fascia and the side of the house.

How to Repair Wood Fascia

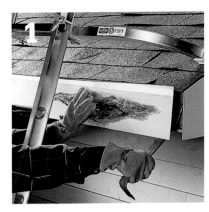

Remove gutters, shingle moldings, and any other items mounted on the fascia. Carefully pry off the damaged fascia board using a pry bar. Remove the entire board and all old nails.

Set your circular saw for a 45° bevel, and cut off the damaged portion of the fascia board. Reattach the undamaged original fascia to the rafters or rafter lookouts using 2" deck screws. Bevel-cut a patch board to replace the damaged section.

Set the patch board in place. Drill pilot holes through both fascia boards into the rafter. Drive nails in the holes to create a locknail joint (inset). Replace shingle moldings and trim pieces using 4d casing nails. Set the nail heads. Prime and paint the new board.

How to Repair Wood Panel Soffits

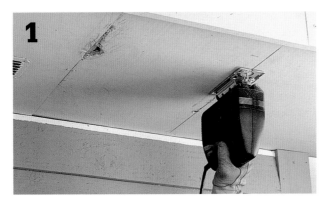

1

In the area where soffits are damaged, remove the support moldings that hold the soffits in place along the fascia and exterior wall. Drill entry holes, then use a jigsaw to cut out the damaged soffit area. *Tip: Cut soffits as close as possible to the rafters or rafter lookouts. Finish cuts with an oscillating tool or a chisel, if necessary.*

2

Remove the damaged soffit section using a pry bar. Cut nailing strips the same length as the exposed area of the rafters, and fasten them to the rafters or rafter lookouts at the edges of the openings using 2½" deck screws.

3

Using soffit material similar to the original panel, cut a replacement piece ⅛" smaller than the opening. If the new panel will be vented, cut the vent openings.

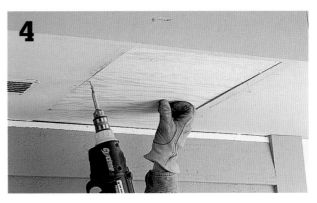

4

Attach the replacement panel to the nailing strips using 2" deck screws. If you are not going to paint the entire soffit after the repair, prime and paint the replacement piece before installing it.

5

Reattach the soffit molding using 4d casing nails. Set the nail heads.

6

Using siliconized acrylic caulk, fill all nail holes, screw holes, and gaps. Smooth out the caulk with a putty knife until the caulk is even with the surface. Prime and paint the soffit panels.

Gutters

Vinyl Gutters

Installing a snap-together vinyl gutter system is a manageable task for most do-it-yourselfers. Snap-together gutter systems are designed for ease of installation, requiring no fasteners other than the screws used to attach the gutter hangers to the fascia.

Before you purchase new gutters, create a detailed plan and cost estimate. Include all of the necessary parts, not just the gutter and drain pipe sections—they make up only part of the total system. Test-fit the pieces on the ground before you begin the actual installation.

Tools & Materials ▸

Chalkline
Tape measure
Drill
Hacksaw
1¼" deck screws
Gutters
Drain pipes

Connectors
Fittings
Hangers

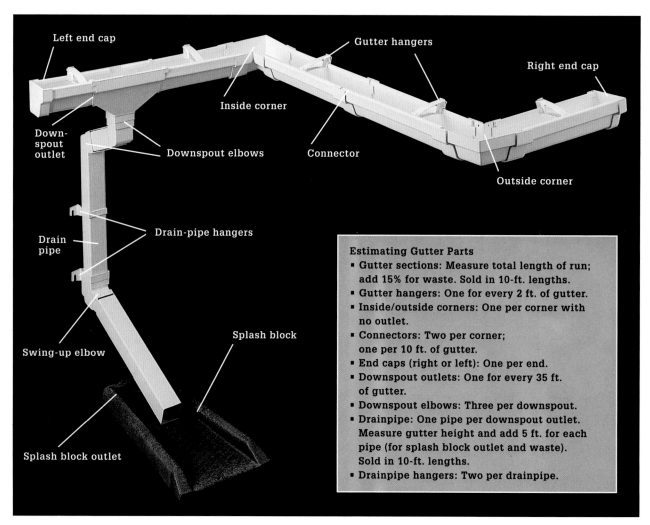

Estimating Gutter Parts
- Gutter sections: Measure total length of run; add 15% for waste. Sold in 10-ft. lengths.
- Gutter hangers: One for every 2 ft. of gutter.
- Inside/outside corners: One per corner with no outlet.
- Connectors: Two per corner; one per 10 ft. of gutter.
- End caps (right or left): One per end.
- Downspout outlets: One for every 35 ft. of gutter.
- Downspout elbows: Three per downspout.
- Drainpipe: One pipe per downspout outlet. Measure gutter height and add 5 ft. for each pipe (for splash block outlet and waste). Sold in 10-ft. lengths.
- Drainpipe hangers: Two per drainpipe.

Vinyl snap-together gutter systems are easy to install and relatively inexpensive, and they won't rot or deteriorate. The slip joints allow for expansion and contraction, which contribute to their reliability and longevity.

How to Install Vinyl Gutters

Mark a point at the high end of each gutter run, 1" from the top of the fascia. Snap chalklines that slope ¼" per 10 ft. toward downspouts. For runs longer than 35 ft., mark a slope from a high point in the center toward downspouts at each end.

Install downspout outlets near the ends of gutter runs (at least one outlet for every 35 ft. of run). The tops of the outlets should be flush with the slope line, and they should align with end caps on the corners of the house.

Following the slope line, attach hangers or support clips for hangers for a complete run. Attach them to the fascia at 24" intervals using deck screws.

Following the slope line, attach outside and inside corners at all corner locations that don't have end caps.

(continued)

Use a hacksaw to cut gutter sections to fit between outlets and corners. Attach the end caps and connect the gutter sections to the outlets. Cut and test-fit gutter sections to fit between outlets, allowing for expansion gaps.

Working on the ground, join the gutter sections together using connectors. Attach gutter hangers to the gutter (for models with support clips mounted on the fascia). Hang the gutters, connecting them to the outlets.

Cut a section of drainpipe to fit between two downspout elbows. One elbow should fit over the tail of the downspout outlet and the other should fit against the wall. Assemble the parts, slip the top elbow onto the outlet, and secure the other to the siding with a drainpipe hanger.

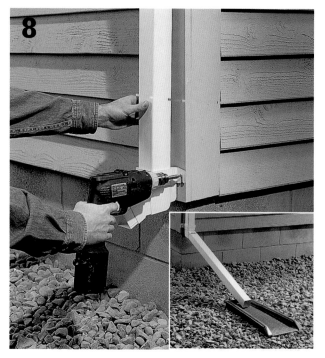

Cut a piece of drainpipe to fit between the elbow at the top of the wall and the end of the drainpipe run, staying at least 12" above the ground. Attach an elbow, and secure the pipe to the wall with a drainpipe hanger. Add accessories, such as splash blocks, to help channel water away from the house (inset).

Metal Gutters

Most home centers carry sections of aluminum or galvanized steel gutter along with the components necessary to install your own system. Gutter sections and downspouts are typically 10 ft. long, though if you order them online and have them shipped to your home they'll arrive in 5-ft. sections. Assembly is easy. Pieces go together with self-tapping screws and sealant, though pop-rivets also prove handy.

Installation is the hard part. You'll be working at heights with long components. While up there you'll need to attach the gutter so it smoothly inclines toward the downspout at a slope of ¼" for every 10 ft. Enlist a helper for installation or rig up a loop of rope to hold one end of the gutter while you attach the hangers to the fascia. Typically, you can use your old gutter setup as a guide for the new.

Tools & Materials ▶

Tape measure
Drill with ¼"
    hex-drive bit
Caulk gun
Chalkline
Hammer
Tin snips
Hacksaw
Gutters
Zip screws
Gutter sealant or
    silicone caulk

Hangers
Gutter outlets
Downspouts
End caps
Elbows
Hangers
Downspout brackets
End box
Speed square

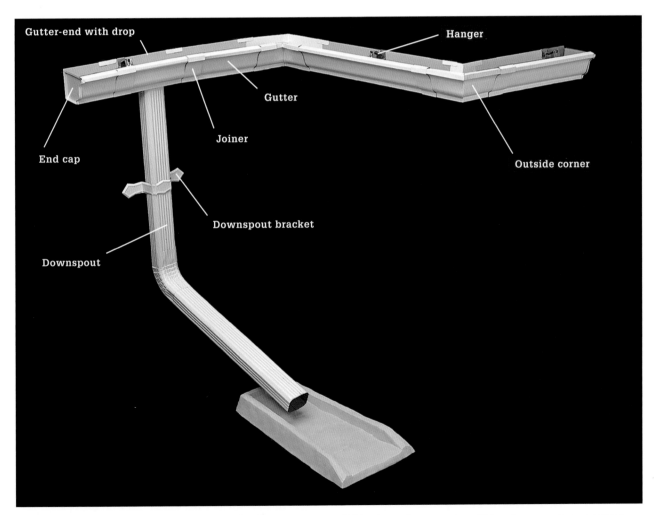

Gutter-end with drop

Hanger

Gutter

End cap

Joiner

Outside corner

Downspout bracket

Downspout

Sealed and fastened correctly, gutters made of components available at your home center will function as well as seamless gutters fabricated on site by a contractor. The trick is getting the slope smooth and accurate.

How to Install Metal Gutters

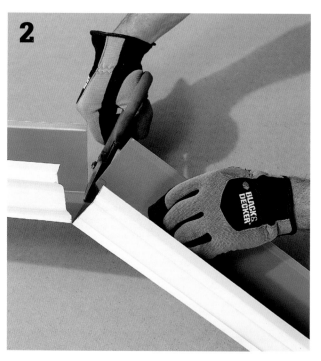

At the fascia's midpoint, measure down from the drip edge and make a mark for the bottom of the gutter. Mark both ends of the fascia, adding a ¼" slope for every 10 ft. of gutter. Snap chalklines between the marks.

Use tin snips to cut the gutter in three steps. After marking all three sides, nip through the top front edge of the gutter, and then work down through the profile. Next, cut the back of the gutter. Finally, let the gutter bend open as shown so you can cut through the bottom.

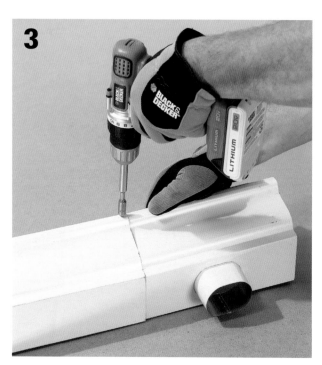

Attach a prefabbed drop to your section of gutter. Apply sealant, join the gutter section to the drop, and attach with zip screws or rivets, whichever the manufacturer recommends.

Place an end cap over the end of the gutter. Drive zip screws through the flange into the gutter. Apply ample sealant along the inside edges of the cap.

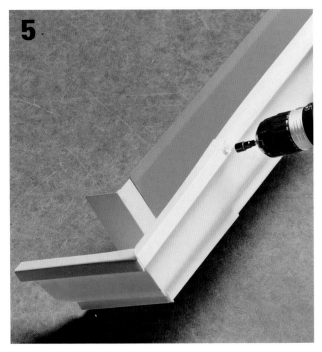

Apply a small bead of sealant on the bottom and sides inside a corner box. Slide the end of the gutter inside the box. Fasten the gutter and box together using zip screws. Apply ample sealant along the inside seam.

Clip gutter hangers to the gutter every 24". Lift the gutter into place, sliding the back side under the drip edge and aligning the bottom with the chalkline. Drive the nail or screw in each hanger through the fascia to install.

Fasten an elbow to the gutter outlet, driving a zip screw through each side. Hold another elbow in place against the house. Measure the distance between the elbows, adding 2" at each end for overlap. Cut a downspout to this length using a hacksaw. Crimp the corners of the downspout for easy insertion and fasten together. *Tip: Assemble the elbows and downspout so the top pieces always fit inside bottom pieces.*

Fasten downspout brackets to the wall for the top and bottom of the downspout and every 8 ft. in between. Cut a downspout that spans the length of the wall, and attach it to the elbow at the top. Install another elbow at the end of the downspout. Fasten the brackets to the downspout.

Repairing Gutters

Gutters perform the important task of channeling water away from your house. A good gutter system prevents damage to your siding, foundation, and landscaping, and it helps prevent water from leaking into your basement. When gutters fail, evaluate the type and extent of damage to select the best repair method. Clean your gutters and downspouts as often as necessary to keep the system working efficiently.

Tools & Materials ▸

Flat pry bar
Hacksaw
Caulk gun
Pop rivet gun
Drill
Hammer
Stiff-bristle brush
Putty knife
Steel wool
Aviation snips
Level
Paintbrush
Trowel
Garden hose
Chalkline
Wood scraps
Replacement
  gutter materials
Siliconized
  acrylic caulk
Roofing cement
Metal flashing
Sheet-metal screws
  or pop rivets
Gutter hangers
Primer and paint
Gutter patching kit
Gutter guards

Use a trowel to clean leaves, twigs, and other debris out of the gutters before starting the repairs.

Keep gutters and downspouts clean so rain falling on the roof is directed well away from the foundation. Nearly all wet basement problems are caused by water collecting near the foundation, a situation that can frequently be traced to clogged and overflowing gutters and downspouts.

How to Unclog Gutters

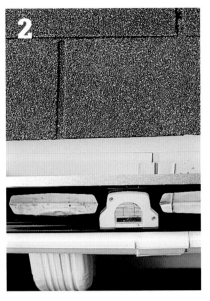

Flush clogged downspouts with water. Wrap a large rag around a garden hose and insert it in the downspout opening. Arrange the rag so it fills the opening, then turn on the water full force.

Check the slope of the gutters using a level. Gutters should slope slightly toward the downspouts. Adjust the hangers, if necessary.

Place gutter guards over the gutters to prevent future clogs.

How to Rehang Sagging Gutters & Patch Leaks

For sagging gutters, snap a chalkline on the fascia that follows the correct slope. Remove hangers in and near the sag. Lift the gutter until it's flush with the chalkline. *Tip: A good slope for gutters is a ¼" drop every 10 ft. toward the downspouts.*

Reattach hangers every 24" and within 12" of seams. Use new hangers, if necessary. Avoid using the original nail holes. Fill small holes and seal minor leaks using gutter caulk.

Use a gutter patching kit to make temporary repairs to a gutter with minor damage. Follow manufacturer's directions.

How to Repair Leaky Joints

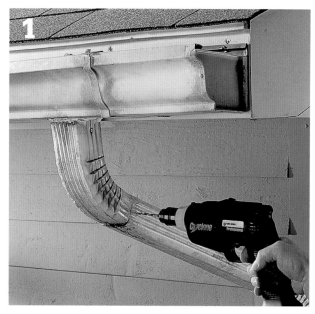

Drill out the rivets or unfasten the metal screws to disassemble the leaky joint. Scrub both parts of the joint with a stiff-bristle brush. Clean the damaged area with water, and allow to dry completely.

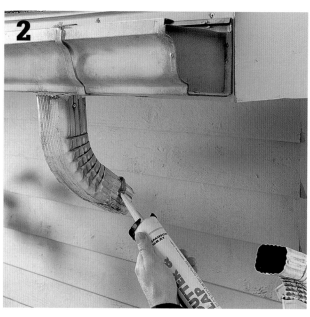

Apply caulk to the joining parts, then reassemble the joint. Secure the connection with pop rivets or sheet-metal screws.

How to Patch Metal Gutters

Clean the area around the damage with a stiff-bristle brush. Scrub it with steel wool or an abrasive pad to loosen residue, then rinse it with water.

Apply a ⅛"-thick layer of roofing cement evenly over the damage. Spread the roofing cement a few inches past the damaged area on all sides.

Cut and bend a piece of flashing to fit inside the gutter. Bed the patch in the roofing cement. Feather out the cement to reduce ridges so it won't cause significant damming. *Tip: To prevent corrosion, make sure the patch is the same type of metal as the gutter.*

How to Replace a Section of Metal Gutter

Remove gutter hangers in and near the damaged area. Insert wood spacers in the gutter near each hanger before prying. *Tip: If the damaged area is more than 2 ft. long, replace the entire section with new material.*

Slip spacers between the gutter and fascia near each end of the damaged area, so you won't damage the roof when cutting the gutter. Cut out the damaged section using a hacksaw.

Cut a new gutter section at least 4" longer than the damaged section.

Clean the cut ends of the old gutter using a wire brush. Caulk the ends, then center the gutter patch over the cutout area and press into the caulk.

Secure the gutter patch with pop rivets or sheet-metal screws. Use at least three fasteners at each joint. On the inside surfaces of the gutter, caulk over the heads of the fasteners.

Reinstall gutter hangers. If necessary, use new hangers but don't use old holes. Prime and paint the patch to match the existing gutter.

Conversions

Metric Conversions

TO CONVERT:	TO:	MULTIPLY BY:
Inches	Millimeters	25.4
Inches	Centimeters	2.54
Feet	Meters	0.305
Yards	Meters	0.914
Square inches	Square centimeters	6.45
Square feet	Square meters	0.093
Square yards	Square meters	0.836
Ounces	Milliliters	30.0
Pints (U.S.)	Liters	0.473 (Imp. 0.568)
Quarts (U.S.)	Liters	0.946 (Imp. 1.136)
Gallons (U.S.)	Liters	3.785 (Imp. 4.546)
Ounces	Grams	28.4
Pounds	Kilograms	0.454

TO CONVERT:	TO:	MULTIPLY BY:
Millimeters	Inches	0.039
Centimeters	Inches	0.394
Meters	Feet	3.28
Meters	Yards	1.09
Square centimeters	Square inches	0.155
Square meters	Square feet	10.8
Square meters	Square yards	1.2
Milliliters	Ounces	.033
Liters	Pints (U.S.)	2.114 (Imp. 1.76)
Liters	Quarts (U.S.)	1.057 (Imp. 0.88)
Liters	Gallons (U.S.)	0.264 (Imp. 0.22)
Grams	Ounces	0.035
Kilograms	Pounds	2.2

Lumber Dimensions

NOMINAL - U.S.	ACTUAL - U.S. (IN INCHES)	METRIC
1 × 2	¾ × 1½	19 × 38 mm
1 × 3	¾ × 2½	19 × 64 mm
1 × 4	¾ × 3½	19 × 89 mm
1 × 5	¾ × 4½	19 × 114 mm
1 × 6	¾ × 5½	19 × 140 mm
1 × 7	¾ × 6¼	19 × 159 mm
1 × 8	¾ × 7¼	19 × 184 mm
1 × 10	¾ × 9¼	19 × 235 mm
1 × 12	¾ × 11¼	19 × 286 mm
2 × 4	1½ × 3½	38 × 89 mm
2 × 6	1½ × 5½	38 × 140 mm
2 × 8	1½ × 7¼	38 × 184 mm
2 × 10	1½ × 9¼	38 × 235 mm
2 × 12	1½ × 11¼	38 × 286 mm
3 × 6	2½ × 5½	64 × 140 mm
4 × 4	3½ × 3½	89 × 89 mm
4 × 6	3½ × 5½	89 × 140 mm

Metric Plywood Panels

Metric plywood panels are commonly available in two sizes: 1,200 mm × 2,400 mm and 1,220 mm × 2,400 mm, which is roughly equivalent to a 4 × 8-ft. sheet. Standard and Select sheathing panels come in standard thicknesses, while Sanded grade panels are available in special thicknesses.

STANDARD SHEATHING GRADE		SANDED GRADE	
7.5 mm	(5⁄16 in.)	6 mm	(4⁄17 in.)
9.5 mm	(3⁄8 in.)	8 mm	(5⁄16 in.)
12.5 mm	(½ in.)	11 mm	(7⁄16 in.)
15.5 mm	(5⁄8 in.)	14 mm	(9⁄16 in.)
18.5 mm	(¾ in.)	17 mm	(2⁄3 in.)
20.5 mm	(13⁄16 in.)	19 mm	(¾ in.)
22.5 mm	(7⁄8 in.)	21 mm	(13⁄16 in.)
25.5 mm	(1 in.)	24 mm	(15⁄16 in.)

Liquid Measurement Equivalents

1 Pint	= 16 Fluid Ounces	= 2 Cups
1 Quart	= 32 Fluid Ounces	= 2 Pints
1 Gallon	= 128 Fluid Ounces	= 4 Quarts

Converting Temperatures

Convert degrees Fahrenheit (F) to degrees Celsius (C) by following this simple formula: Subtract 32 from the Fahrenheit temperature reading. Then, multiply that number by $\frac{5}{9}$. For example, 77°F - 32 = 45. 45 × $\frac{5}{9}$ = 25°C.

To convert degrees Celsius to degrees Fahrenheit, multiply the Celsius temperature reading by $\frac{9}{5}$. Then, add 32. For example, 25°C × $\frac{9}{5}$ = 45. 45 + 32 = 77°F.

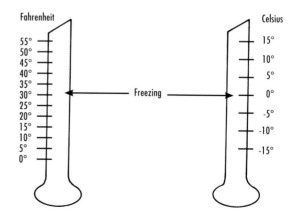

Drill Bit Guide

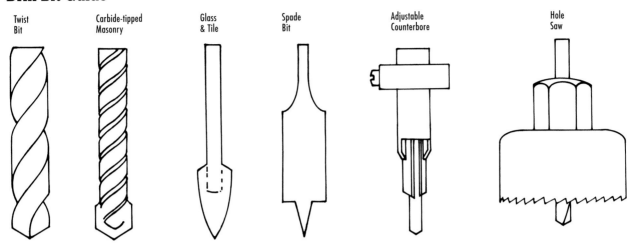

Twist Bit   Carbide-tipped Masonry   Glass & Tile   Spade Bit   Adjustable Counterbore   Hole Saw

Nails

Nail lengths are identified by numbers from 4 to 60 followed by the letter "d," which stands for "penny." For general framing and repair work, use common or box nails. Common nails are best suited to framing work where strength is important. Box nails are smaller in diameter than common nails, which makes them easier to drive and less likely to split wood. Use box nails for light work and thin materials. Most common and box nails have a cement or vinyl coating that improves their holding power.

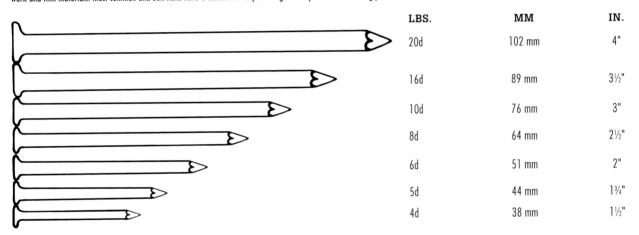

LBS.	MM	IN.
20d	102 mm	4"
16d	89 mm	3½"
10d	76 mm	3"
8d	64 mm	2½"
6d	51 mm	2"
5d	44 mm	1¾"
4d	38 mm	1½"

Counterbore, Shank & Pilot Hole Diameters

SCREW SIZE	COUNTERBORE DIAMETER FOR SCREW HEAD (IN INCHES)	CLEARANCE HOLE FOR SCREW SHANK (IN INCHES)	PILOT HOLE DIAMETER	
			HARD WOOD (IN INCHES)	SOFT WOOD (IN INCHES)
#1	9/64	5/64	3/64	1/32
#2	1/4	3/32	3/64	1/32
#3	1/4	7/64	1/16	3/64
#4	1/4	1/8	1/16	3/64
#5	1/4	1/8	5/64	1/16
#6	5/16	9/64	3/32	5/64
#7	5/16	5/32	3/32	5/64
#8	3/8	11/64	1/8	3/32
#9	3/8	11/64	1/8	3/32
#10	3/8	3/16	1/8	7/64
#11	1/2	3/16	5/32	9/64
#12	1/2	7/32	9/64	1/8

Resources

Alside
800-922-6009
www.alside.com

Black & Decker
Portable Power Tools & More
www.blackanddecker.com

Boral Roofing Company
909-737-0200
www.boral.com

CertainTeed, Corp.
800-782-8777
www.certainteed.com

DaVinci Roofscapes
Faux slate shingles, page 86
855-299-5301
www.davinciroofscapes.com

GAF Materials, Corp.
877-423-7663
www.gaf.com

Gentite
877-436-8483
www.gentite.com

Interlock
888-766-3661
www.interlockroofing.com

James Hardie© Siding Products
888-542-7343
www.jameshardie.com

La Habra Stucco
800-226-2424
www.lahabrastucco.com

MCA Superior Clay Roof Tile
800-736-6221
www.mca-tile.com

Midwest Lumber
800-862-6003
www.midwestlumberinc.com

National Roofing Contractors Association (NRCA)
847-299-9070
www.nrca.net

Novabrik
866-678-2745
www.novabrik.com

Plygem
800-788-1964
www.plygem.com

Red Wing Shoes
Work shoes and boots shown
    throughout book
800-733-9464
www.redwingshoes.com

Rooflite
www.rooflite.com

The Tapco Group
800-521-7567
www.tapcoint.com

US Tile
800.252.9548
www.ustile.com

Photo Credits

Alamy
page 124 (left)

Alcoa Home Exteriors, Inc.
pages 18, 119, 127, 138

Alside
pages 46, 116 (main), 234

Andrea Rugg
pages 24, 117

Boral Roofing Company
pages 15 (top), 23

CertainTeed, Corp.
pages 12, 24, 28 (lower left), 115, 118,
    121, 146

DaVinci Roofscapes
pages 15 (bottom), 86

Dreamstime.com
pages 27, 62, 162, 226

EcoStar, LLC
page 26

FR Midwest
page 170

GAF Materials, Corp.
pages 10, 14, 19, 21, 44

Gentite
pages 20, 74

Interlock
pages 25 (top left, top right), 78

Istockphoto.com
page 112

James Hardie
page 166

La Habra Stucco
page 192

Novabrik
pages 125 (bottom), 184

Scot Zimmerman
pages 122 (left), 236 (left)

Shutterstock.com
page 79

The Tapco Group
page 120

US Tile
pages 22, 93, 97

Index

A

air nailers, 31
alligatoring paint, 202
aluminum end caps, 221
aluminum fascia, 234–35
aluminum siding. *See* metal siding
aluminum soffits, 230–33
architectural asphalt shingles. *See*
  laminated asphalt shingles
asphalt roll roofing, 19
asphalt shingles
  asphalt three-tab shingles, 16–17,
    51–57
  cleaning, 109
  costs of, 14, 16
  damaged, 98
  durability, 15–16
  installation, 16, 18, 51–57
  laminated asphalt shingles, 18
  lifting and staging, 34
  maintenance, 99
  repairs, 101
  replacement of, 15, 102
  ridge vents, 60–61
  shingling over old roofs, 58–59
  tools and materials for, 50
  types of, 14, 16–17
asphalt three-tab shingles, 16–17, 51–57

B

back support braces, 34
base flashing, 36–37
bending jigs, 48–49
bleeding spots, 203
blistering paint, 202–3, 218
board and batten siding
  about, 119, 122
  costs of, 114
  installation, 167–69
  removal of, 135
  repairing, 222
  tools and materials for, 166
brickmold, 135, 176–77
brick siding
  about, 124, 180
  costs of, 114
  installation, 180–83
  maintenance, 115
  mortarless brick veneer, 184–89
  roofing style and, 14
buckling shingles, 98
building codes, 24, 56, 58, 62, 180
building paper. *See* underlayment

C

cast veneer stone, 191
cedar shakes/shingles
  about, 62–63

costs for, 14
durability, 15
installation, 64–67
replacement of, 15, 103
underlayment and, 63
ceiling stains, 13
cement fiber siding
  about, 118, 121, 125
  costs of, 114
  cutting of lap, 152
  installation, 114, 150–55
  maintenance of, 115
cement tile roofing, 12
chalk, 51
chimney cap, 37
chimneys, 55–56
clapboard siding. *See* wood lap siding
clay tile roofing
  about, 92
  costs for, 14, 22
  durability, 15, 22
  installation, 22, 94–97
  repairs of, 22
  style considerations and, 12, 14, 22
  weight of, 22
climate
  asphalt three-tab shingles and, 16–17
  roofing selection and, 15
  underlayment and, 44
  ventilation and, 228
  zinc strips and, 99
clothing, 30
concrete pavers, 20
concrete tile roofing
  costs for, 14, 25
  durability, 15, 23
  installation, 23
  weight of, 23
copper roofing, 14
corrugated nonmetallic panels, 25. *See
  also* metal roofing
counter flashing, 36–37
cupping shingles, 13, 98

D

dimensional asphalt shingles. *See*
  laminated asphalt shingles
doors, 217
dormers, 52–55, 229
drip edge, 46–47, 80
dumpsters, 40

E

efflorescence, 203
electricity, 31
environmentally friendly roofing, 14,
  26–27, 50
EPDM rubber roofing, 20, 75–77

equipment. *See* tools and materials
ethylene propylene diene monomer
  (EPDM) roofing. *See* EPDM
  rubber roofing

F

fall-arresting gear, 30, 33
fascia
  aluminum, 234–35
  painting, 211
  in roof systems, 36–37
  wood, 238–39
faux slate roofing
  about, 86–87
  installation, 88–91
felt paper. *See* underlayment
fiber cement roofing, 23. *See also*
  concrete tile roofing
fiber cement siding. *See* cement
  fiber siding
fiberglass asphalt roofing, 14
flashing
  base flashing, 36–37
  bending, 48–49
  chimneys and, 55–56
  clay tile roofing, 94–97
  counter flashing, 36–37
  deterioration of, 13
  drip edge, 46–47, 80
  estimation, 28
  faux slate and, 89–91
  gutter repair and, 248
  installation, 49
  loosening of, 13, 15
  raised seam metal roofing and,
    80–85
  rake flashing, 83–84
  repairing, 101
  roofing cement and, 48
  roof systems, 36–37
  shingling over old roofs, 58–59
  slate roofing and, 21
  step flashing, 108
  types of, 36–37, 39, 48
  valley flashing, 48–49, 104–7
  vent pipes, 48, 53–54
  wall flashing, 84

G

gables, 229
GFCI extension cords, 31
green roofing, 14, 26–27, 50
gutters
  metal, 243–45
  repairing, 246–49
  in roof systems, 36–37
  siding maintenance and, 115
  vinyl, 240–42

H

helpers
  safety and, 30–31
  staging shingles and, 34
  time estimation and, 29
housewrap
  installation, 136–37

I

ice dams
  inspection of, 98
  roof sheathing and, 42
  ventilation and, 228, 230
ice guard. *See* underlayment

L

ladder jacks, 126
ladders, 30–32
laminated asphalt shingles, 18
landscaping
  siding maintenance and, 115
lap siding
  about, 119
  cleaning, 213
  costs of, 114
  cutting, 147–48, 152
  damaged panels, 218
  installation, 149–55
  maintenance of, 115
  removal of, 135
  repairing, 223
  separated joints, 219
  tools and materials for, 146–48
leadership in energy and
  environmental design (LEED)
  credits, 26
leaf removal, 15
leaks
  ceiling stains and, 13
  evaluation of, 100
  gutters and, 247–48
  repairing, 100
  roof sheathing and, 42
living roofs, 27
log cabin siding
  about, 123
  installation, 171–75
  tools and materials for, 170
lumber support braces, 34

M

masonry siding, 178–79. *See also* brick
  siding; stone siding; stucco siding
metal gutters
  about, 243
  installing, 244–45
  repairing, 246–49
metal roofing
  climate and, 24
  costs of, 14
  durability, 15, 24–25
  installation, 24–25, 80–85

raised seam overview, 78
shingling over old roofs and, 78
style considerations and, 14, 24
weight of, 24
metal shingles, 25. *See also*
  metal roofing
metal siding
  about, 116–17
  damaged siding, 219
  paint removal, 207
  repairing, 220–21
mildew, 99, 203
mortarless brick veneer siding, 184–89
mortarless stone veneer siding,
  124–25
moss, 15, 99
multithickness asphalt shingles. *See*
  laminated asphalt shingles

O

organic asphalt roofing, 14, 50

P

paint brushes and rollers, 209–10, 212
painting
  brickmold, 176
  brushes and rollers, 209–10, 212
  cleaning tools, 212
  estimating paint needs, 201
  paint application, 208, 210–12
  paint problems, 202–3
  paint removal, 201, 204–7
  paint sprayers, 214–17
  preparing surfaces for, 204–7
  tools and materials for, 200–201
paint problems, 202–3
paint sprayers, 214–17
panel siding, 121
peeling paint, 202–3, 218
photovoltaic shingles, 27
plastic tarps, 14
plywood siding
  costs of, 114
  installation, 114, 163–65
  tools and materials for, 162
pneumatic nailers, 31
premixed stucco products, 194
pressure washing, 99
pump-jack scaffolding, 132–33

R

raised seam metal roofing
  installation, 80–85
  overview of, 78
rake flashing, 83–84
ridge vents, 60–61, 229
roll roofing
  about, 68
  asphalt roll roofing, 19
  costs of, 14, 19–20
  durability, 15, 19
  EPDM rubber roofing, 20, 74–77

installation, 19, 69–73
repairing, 15
self-adhesive, 14, 19, 72–73
roofing
  choosing, 14–15
  cleaning, 109
  colors, 12, 14
  costs and, 11–12, 14
  durability and, 15
  emergency repairs, 101
  estimating, 28
  inspecting, 98–100
  lifting/staging shingles and, 34
  maintenance, 99, 109
  problems, 13, 98–100
  ridge vents, 60–61
  roof jacks, 35
  roof systems, 36–37
  sheathing, 36–37, 42–43
  shingling over old roofs, 58–59, 78
  tear off, 41–42
  time estimation, 29
  tools and materials for, 38–39
  underlayment, 39, 44–45
  See also individual types
roofing cement, 19, 48, 101, 248
roofing disposal, 14
roofing ladders, 38
roofing problems, 13, 98–100
roof jacks, 29, 35, 38
roof sheathing
  replacing, 42–43
  in roof systems, 36–37
roof slope
  asphalt shingles and, 16, 18
  calculation of, 29
  EPDM rubber roofing and, 20
  fall-arresting gear and, 30
  living roofs and, 27
  roofing estimation and, 28
  roofing selection and, 15
roof vents
  in roof systems, 36–37
  ventilation and, 228–29
rubber roofing. *See* EPDM
  rubber roofing
rust, 203, 207

S

safety
  electricity and, 31
  fall-arresting gear and, 30, 33
  helpers and, 30–31
  ladders and, 30–32
  roofing and, 11, 30–33
  roof jacks and, 35
  roof sheathing and, 42
  staging shingles and, 34
  weather and, 30
sagging ridges, 99
saw areas, 129

scaffolding
    pump-jack scaffolding, 132–33
    setting up, 130–31
self-adhesive roll roofing, 14, 19, 72–73
sheathing. *See* roof sheathing
shiplap siding. *See* wood lap siding
shutters, 128
siding
    brickmold and, 176–77
    choosing, 113–14
    estimating, 127
    evaluating, 218–19
    housewrap, 136–37
    maintenance, 115
    painting, 208–12
    preparing walls for stucco, 195
    removal of, 134–35
    repairing, 220–25
    scaffolding, 130–33
    work site preparation, 126, 128–29
    See also individual types
skylights, 66
slate roofing
    costs of, 14, 21
    durability, 15, 21
    faux slate roofing, 86–91
    installation, 21
    replacement of, 21
    style considerations and, 14
    weight of, 21
soffit panels, 36–37
soffits
    aluminum, 230–33
    painting, 211
    roof systems, 36–37
    ventilation and, 228–29
    wood, 236–39
soffit vents, 36–37, 228–29
standing-seam metal roofing, 12, 24.
    See also metal roofing
standing water, 15, 29
step flashing, 108
stone siding, 114–15, 190–91
stucco siding
    about, 124, 192
    costs of, 114
    finishing walls, 196–97
    maintenance, 115
    painting, 212
    premixed stucco products, 194
    preparing walls, 195
    removal of, 135
    repairing, 224–25
    roofing styles and, 14
    stucco systems, 193
    surface bonding cement, 198–99
surface bonding cement, 198–99

T
tar paper. *See* underlayment
tearoff
    equipment and instructions, 40–41
    metal roofing and, 24
    preparation for, 29

tool platforms, 128
tools and materials
    aluminum fascia, 234
    aluminum soffits, 230
    asphalt shingle installation, 50
    board and batten siding, 166
    brickmold, 176–77
    brick siding, 180
    cast veneer stone, 190
    cedar shakes/shingles, 62
    clay tile roofing, 92
    cleaning wood siding, 213
    drip edge, 46
    EPDM rubber roofing, 75–76
    fall-arresting gear, 33
    faux slate roofing, 87
    gutter repair, 246
    log cabin siding, 170
    masonry siding, 178–79
    metal gutters, 243
    mortarless brick veneer, 184
    painting, 200–201, 208–10, 212
    painting preparation, 204
    paint sprayer use, 214
    plywood siding, 162
    raised seam metal roofing, 78
    repairing/inspecting roofs, 98
    ridge vents, 60
    roll roofing, 68
    roof cleaning, 109
    roofing, 38–39
    roof jacks, 35
    scaffolding, 130, 132
    sheathing replacement, 42–43
    shingling over old roofs, 58–59
    siding removal, 134
    siding repair, 220, 224
    stucco siding, 192, 224
    surface bonding cement, 198
    tear off, 40
    underlayment, 44
    valley flashing replacement, 104
    vinyl gutters, 240
    vinyl siding, 138–39
    wood lap siding, 146–48
    wood shake/shingle siding, 156
    wood soffits, 236
tree limbs, 99
trim
    brickmold, 135, 176–77
    evaluating, 218
    lap siding and, 150–52
    paint brushes, 209
    painting, 211
    paint removal, 204–6
    paint removal of metal, 207
tuckpointing, 115

U
underlayment
    asphalt shingle installation, 51
    cedar shakes/shingles, 62–63
    installation, 45
    tools for, 39, 44

V
valley flashing
    installation, 48–49
    raised seam metal roofing and,
        80–81
    replacing, 104–7
    in roof systems, 36–37
ventilation, 228–29
vent pipes
    asphalt shingle installation, 53–54
    raised seam metal roofing and,
        82–83
    roll roofing and, 70
    in roof systems, 36–37, 48
vinyl gutters
    about, 240
    installation, 241–42
    repairing, 246–49
vinyl shakes/shingles, 15
vinyl siding
    about, 116–17
    buckling, 219
    costs of, 114
    cutting, 139
    installation, 140–45
    maintenance, 115
    removal of, 135
    repairing, 220–21
    tools and materials for, 138–39

W
wall flashing, 84
water damage, 98, 100
weather, 30, 99
wood fascia, 238–39
wood shake siding
    about, 120
    clear finish removal, 206
    installation, 157–61
    maintenance, 115
    missing siding, 219
    removal of, 135
    repairing, 222
    tools and materials for, 156
wood shakes/shingles. *See* cedar
    shakes/shingles
wood shingle siding
    about, 120
    clear finish removal, 206
    costs of, 114
    installation, 114, 157–61
    maintenance, 115
    missing siding, 219
    removal of, 135
    repairing, 222
    tools and materials for, 156
wood soffits, 236–39

Z
zinc strips, 99